William Pratt Wainwright

Radical-Mechanics of Animal Locomotion

With Remarks on the Setting-Up of Soldiers, Horse and Foot....

William Pratt Wainwright

Radical-Mechanics of Animal Locomotion
With Remarks on the Setting-Up of Soldiers, Horse and Foot....

ISBN/EAN: 9783337113087

Printed in Europe, USA, Canada, Australia, Japan

Cover: Foto ©berggeist007 / pixelio.de

More available books at **www.hansebooks.com**

RADICAL-MECHANICS

OF

ANIMAL LOCOMOTION.

WITH REMARKS

ON THE

SETTING-UP OF SOLDIERS,

HORSE AND FOOT,

AND ON THE

SUPPLING OF CAVALRY HORSES.

BY

WILLIAM PRATT WAINWRIGHT,

FORMERLY COLONEL COMMANDING SEVENTY-SIXTH NEW YORK
INFANTRY-VOLUNTEERS.

NEW YORK:
PUBLISHED FOR THE AUTHOR BY
D. VAN NOSTRAND,
23 MURRAY AND 27 WARREN STREETS.

1880.

Entered according to Act of Congress, in the year 1880,
By WILLIAM PRATT WAINWRIGHT,
In the office of the Librarian of Congress, at Washington, D. C.

TO THE

Seventy-Sixth New York Infantry-Volunteers,

WHOM I HAD THE HONOR OF COMMANDING

FROM JULY, 1862, TO JUNE, 1863,

INCLUDING THE ACTION OF GAINESVILLE, THE BATTLE OF
SOUTH MOUNTAIN, AND OTHER ENGAGEMENTS,

THESE PAGES

ARE MOST RESPECTFULLY DEDICATED.

PREFACE.

The following discussion, if it establish a method of accomplishing the ends described in its title, may be of use beyond the sphere of contributing to the "setting-up" of soldiers; for "being set-up," in its proper sense, should give additional strength and activity to any man. In fact, setting-up aims at the restoration of the human frame, so far as locomotion and position are concerned, to the perfection in which God created it; a derangement of this perfection being the real obstacle which hinders a man's body from perfectly following the motion of a horse, when riding, or from accommodating its balance to any normal posture in which it may be placed.

If setting-up were applicable only to soldiers, it were well worthy of study, for the calling of a soldier, although liable to great abuse—as is any powerful instrument when used by wicked hands—is, in itself, a high one, and the Holy Scriptures continually mention the noble traits which it requires. Whether it be a Christian calling depends entirely upon the use made of it, and few will question the assumption that the militia, as on a grand scale, the police of the country, and on an emergency, its defenders, are fulfilling a religious as well as a civil duty, in fitting themselves for these objects.

Now, although thorough military discipline is the grand strength of an army, and the indispensable requisite for the success of small numbers against great ones; yet, setting-up, which so far as his body goes, enables a man to look and feel like a soldier, is the physical beginning of this discipline, and with soldiers, as with children, the physical education has great influence on the mental. Beside this, a well set-up man may complete his education as a private after the army is in motion; but setting-up can hardly be accomplished when the drill ground is once abandoned.

In the following discussion we have taken the ground that only a man who is ambidexter can have the perfect command and full force of the movements of his body; whether, indeed, some slight inclination to one side be necessary to avoid a sort of *dead-centre* catch, we cannot positively say, since some observers affirm that they have discovered the tendency to a favorite side, even in wild animals; and also since it is obvious that the position of the stomach, subject as this organ is, to an increase in size disproportionate to its pendant the liver, would favor a right hand and left leg preponderance which, indeed, is so general among civilized men that it has come to be considered as the normal condition of the frame. However this may be, we think it evident that the preponderance need be but very slight, and that anything beyond this measure interferes with the force and free movement of both sets of diagonal limbs.

The fundamental action, as we have in the following pages traced the theory of locomotion, is a helical turning of the appuis, discharged and reduced by a contrary helical turn, whose SUDDEN *discharge constitutes a spring.* The chief appui may be single, as in the fish, where the ribs and fins are mere accessories, and the back-bone acting on the tail gives propulsion to the animal. Or it may be double and have independent working forces, while still directed and to some extent worked by the simple spine, which then becomes a neck, as in the bird, where the wings and legs are attached to a body whose vertebræ are consolidated into a single spinal piece.

Or, this final vertebra may be resolved into a series of vertebræ separating the fore-and rear-limbs, in which case the ribs move on an artificial ground furnished by the breast-bone. The trunk, then, forming a compound spine, works on four limbs, exterior to the whole, as in quadrupeds.

Or, bringing the centre of motion more forward, and more between the fore limbs, the animal may be able to advance to some advantage with the two rear limbs only as appuis on the ground.

Or, finally, the centre of motion being brought still further up, so that the neck, working between the shoulders and the lower jaw, becomes the master-centre of all motion, we have that highest development of vertebrate structure into which alone it has pleased God, as the consummation of His plan for terrestrial creatures, in addition to intelligence or even reason, to breathe a "living soul" capable of knowing its Creator.

We think the epochs between these classes are widely marked, and we are chiefly puzzled as to where one may place the snake, which seems rather to have come from stripping some animal of its limbs than as a regular forward step; for while the fish may contain the elements for fashioning a bird, and the bird for fashioning a quadruped, the snake, if put in the series, must stand below the fish, which has additional appuis, although the snake is evidently of a more perfect organization.

Leaving this point, which we are not sufficiently versed in comparative anatomy to debate, we will only add that the snake, furnishing the most purely simple method of locomotion among the vertebrate animals, affords on that account the clearest ground for tracing the elementary motions.

Our object is not only to propose a theory, but in case this theory be the true one, to make it popularly intelligible. This object, and want of skill in composition will, we hope, excuse considerable prolixity where a good writer could have submitted his views to the decision of qualified judges in much fewer words.

Two articles published in the August and October, 1853, numbers of the *Eclaireur*, a military journal issued for some time by Colonel Cowman, and afterward by General J. Watts de Peyster, and one or two allusions to the subject possibly made in a series of articles in the *Army and Navy Journal*, on " Marching of troops in large bodies," July 2d, 1864, and following ; " The discipline and care of troops," October 1st, 1864, and following; *and* " The fighting of troops," February 11th, 1865, and following; are all that the author has previously written on the subject.

SUMMARY.

The excessive use of one hand, and of the parts of the body brought into action with it, is the cause of a general deformity among civilized men.

This so interferes with the central-pivot working of the body as to greatly reduce its power of producing and sustaining action.

The working of the spine is the fundamental basis of movement.

Motion—properly—originates in the spine, is directed by the head, and is only followed up by the limbs.

The snake presents the simplest type of the spinal working.

Exemplification of the snake's movement by twisting a cord or elastic rod by counter-turning its two ends.

These counter-turns, which produce curvatures similar to those of the snake in locomotion, will, when carried beyond a certain limit, originate from their central point of counteraction reverse curvatures, which, if allowed to replace the old ones and again to produce a fresh set, would present the shapes of alternately reversed curvatures as they are seen in the locomotion of the snake.

In displacing one set of curvatures by their alternates a spring is produced.

The reverse set of curvatures developed from the central point of counteraction of the opposing turns belongs to the nascent alternate curve, but wait to be accepted as such until, by a change of originating points, the old curvatures are discharged and the alternate ones adopted.

The first effect of twisting one of the ends of the elastic rod is to develope a general winding line of shape from one end to the other.

So soon as the twist from one end is fully resisted by that from the other, the *line of counteraction across the thickness of the rod* has that end of it which the active turn directly affects drawn to one side.

The drawing to one side of this end fixes the other end of the cross line which is affected by the counter-twist so that this latter cannot work directly upon it from the originating point.

This fixed displacement of one end of the cross line causes the working of the second twist to be diffused through the rod, and thus the ensuing spring may have for its points of appui a point at the centre of the rod, and a point of rest on the ground at one end of the rod, and on the same side as that at the centre, instead of having both points at the centre and on opposite sides.

It is thus that a movement in progression or retrogression may be obtained, instead of a spring in two directions, from the centre of the rod.

The fundamental action of locomotion is, then, the formation and discharge of two counter-turns in an elastic rod; the turn discharged giving the spring, and the other, after controlling this spring, presenting through the counter-turn it has developed at the cross line, the shape for the alternate of the first, which again receives a counter-turn, and so on.

The displacement to a greater or less degree of the end of the cross line affected by the second turn of one set of diagonal counter-turns becoming permanent, and thus interfering with the action of the alternate set is the cause of the general deformity spoken of above. The reduction of this displacement by strengthening those muscles which work the alternate winding lines, and the giving position to the alternate and opposing cross line, is its cure.

The body of the snake presents an elastic compound rod, which moves by forming, discharging, and, after each spring, replacing by opposite curvatures the twists formed in its back bone and in its ribs.

These twists are projected in a consolidated action in the ball and socket joint of the head, and in partial concentrations in the various ball and socket joints connecting the ribs and the spine.

The simplest example of the action is in any one of the snake's ribs, which being in appui on the ground is, by the motion of the spine, turned through its length in one direction; then, being still fixed at the ground, it is by another motion of the spine turned in the opposite direction, and the spring is liberated by the discharge of the first turn at the ball and socket joints by which the rib joins the spine.

The eyes are the centre of direction, and in the snake as in other animals, are kept steady by the arrangement of a principal muscle passing through a loop, in which, after receiving the counter twists, the two portions into which the loop divides the muscle, slip back to their original condition.

The lung of the snake filled with air presents a permanent compensating-portion of the machinery of locomotion. This, when compressed on one line at the discharge, expands immediately on the alternate line, and thus, having aided in giving force to the discharge of one spring, aids also in the formation of the next. The working of a tense fluid, as the pivot of every movement, also secures the smoothness of action, which, as well as its force, depends much on the filling of the lung.

The snake's ribs, when acting along the several convex and concave curvatures induced by the twistings of the spine, may, on each of them, be divided into Propellers and Bearers, both of which discharge with the primarily formed lines of torsion, but in opposite directions. The propellers, which by a peculiarity in the shape of the curvature, are in preponderance at the rear, thrust the body forward over the bearers, which raise it at the front of the curvature.

All the tractions gather in diagonal action, but discharge in collateral action. This collateral action is only momentary in the movements produced by alternate diagonal counteractions, but if the two possible sets of counteractions are, suppose, to work simultaneously, they reciprocally suppress the diagonal elements of shape, and the alternate helical forms are replaced by undulations in the perpendicular plane.

In these undulations the *force* of the diagonal elements is yet present, but the acting tractions are made collateral even before the discharge.

It is this undulating shape which the spine of the higher animals assumes, more or less, for all springs from two hind feet at once, and its perfect production is required for a perfect halt from any kind of movement.

The position of such a halt is that in which the body is perfectly gathered for making, on the instant, any movement of which it is capable, and is what is meant by "setting-up" when applied to an attitude of immobility.

This position may be produced either by forcing one of the sets of counteractions beyond its limits without discharge, in which case the other set accepts the excess of gathering, and both gatherings thus become formed; or by commencing with the formation of both sets at the same time.

The movements of the snake are caused by a succession of springs from an indefinite number of what may be called torsion curves, of two curvatures each; but in man and the higher animals the spinal curves are limited to three:—one in the back, which answers to the chest curve in the snake; one in the loins, which is not complete until by moving the hinder limbs it has compounded itself with the chest curve; and one in the neck, which combines itself with the other two, and by moving the fore-limbs and lower jaw on a virtual ball and socket joint at the root of the neck, forms the whole body into one curve.

In this doubly compounded curve the hind and fore-limbs are the appuis, the head the director, and the centre of the lungs, made one in action with the head joints is the centre of force.

In man, by means of the collar bones and general shape, such a perfect transmission to the feet of the appui on the arms is effected that his natural position becomes one of sole acutual appui on these latter, although a proper action of the arms as a brace is still a condition of its perfection.

In the higher animals the effects of the lines of torsion forming in the central or ophidian curve pass to the breast bone, which, as an artificial ground, supports the play of the ribs; the effects of the lines of torsion forming in the combination of the central and loin curves, and of a partial movement in the neck, pass to the pelvis and to the uppermost part of the breast bone; and the effects of the lines of torsion forming in the combination of the neck curve with the two others, making one curve of the whole, pass to the limbs, and finally to the front limbs and lower jaw.

To sustain the complicated movements of the higher animals, the construction of the head joint is altered from that of the snake, so that in addition to the general ball and socket movement there is a superimposed movement of the same kind, but divided into two parts, which belongs especially to the limbs and by which the neck concentrates the whole upon its own curve, with the shoulder-blades and lower jaw as bases of action, and, as was before said, with the centre of the lungs and the head joint conjoined in action as a focus of force.

The lungs also are so constituted that they act for the central curve in four parts, the two lower of which belong particularly to the lower sections of the torsion lines whose point of origination is above, and the two upper particularly to the upper sections of those lines whose originating point is below.

The addition of limbs to the locomotive frame-work makes necessary the addition of some comparatively small appendages to the top of the lungs, which fill in a contrary direction to the filling of the lobe to which they are attached. These appendages sustain the ball and socket action at the root of the neck, and their action in filling may, like that of the lungs proper, be divided into four parts.

Two small neck muscles, attached each by one end to the head and by the other to the lower jaw, hold the lungs suspended by a loop on each side of the windpipe. The action of these loops, in releasing the muscles from the twists they receive by the torsion of the lungs in the movements of the chest, is similar to the action of the loops of the eye muscles, before spoken of, in releasing the torsions passing to the eyes from the whole body. These muscles and the eye-muscles together have as analogues, but in subordinate working, the diaphragm, its pillars and the psoæ muscles.

The double twist action is the same whether expanded in its workings, as in the motions through the length of the ribs, or condensed, as in the ball and socket joint of the head.

The various combinations of the lines of torsion in the three curves are the foundation for all the gaits of animals, and for their halting, which last, as before said, when perfectly carried out, is identical with the act of " setting-up;" for the deformity spoken of is really a position in locomotion, and only a deformity because of its being permament ; and even were there no deformity the frame would need to be gathered, which involves the same movements as does halting on two sets of torsion lines simultaneously.

The actions of those winding lines which work from the direction of the head toward the hinder limbs centre at first on the lower jaw; while the action of those which work from the direction of the hinder limbs toward the head centre at first on the shoulder-blades. They interchange from one of these gathering points to the other, and finally, by the movement of the winding-lines in the neck, and the virtual ball and socket at the neck-root, the control of all the forces becomes collected in the neck, and guided by the head working on the shoulders and lower jaw, with the eyes as poles of direction.

By a continuous raising of the head, allowing the neck to adjust itself to the ensuing workings of the body, without allowing any line to come to a discharge, the counteractions of the various lines may all be brought into gathering, and thus afterward require only to be loosened in a particular connection to bring on the appropriate movements.

The application of setting-up to the horse is called in Manege language "Suppling." It may be carried out on precisely the same principles in the horse as in man.

Many years ago a Frenchman, by the name of *Morquin*, taught, in New York, a method of preparation for gymnastic exercises, in which, by forcing the filling of the lungs and thus bringing on the movements of other parts, a complete "setting up" was effected. This method may, we think, be explained by one of the results of the theory here given, but Mons. Morquin could render little or no account of the why and wherefore. This practice seemed to have originated in some accidental discovery when carrying out the idea that the lungs must be well filled as a basis of action.

The methods for setting-up on horseback are, *with allowances for changed appuis*, the same as for setting-up on foot. If the man can give proper action to the cross line which is alternate to that whose winding lines have been unnaturally confirmed, there then remains nothing to prevent his following fully in his own body the gatherings and discharges in the body of his horse, so fully that there will be no inharmonious movement which shall cause him to leave the saddle in any degree.

The management of the horse consists in controlling the lines of counteraction formed in the animals body. The aids to this are :—the Bit; the Spur; the Pressure of the Rein against the neck; and the weighting one or the other Seat-bone of the rider. These bring into action or check different parts of the winding-lines, and by weakening the working of their action or introducing the alternates upon them, enable the rider not only to signify his wishes to the horse but to force compliance.

A short section is added on locomotion in fishes and in birds.

REMARKS
ON THE
SETTING-UP OF SOLDIERS,
HORSE AND FOOT,
AND ON THE
SUPPLING OF CAVALRY HORSES.

INTRODUCTION.

§ 1. Many are the expedients which, in the training of soldiers, have been and still are adopted, in order to overcome that fault in the body, whatever it may be, which, in nine hundred and ninety-nine men out of every thousand from civilized nations, tends to hinder the man from marching in a straight line, from discharging his musket without destroying his aim, from cutting perpendicularly with the edge of his sabre, and which likewise hinders him from so following in his own frame the motions received from the frame of his horse that the forces communicated by this latter shall be so absorbed into and discharged with the working of his own as to give no recoil from the saddle.

From the expedient of carrying "the left eye

over the line of the vest buttons," in Frederick the Great's time, to that of "fixing the chin and pointing the toes" of the present day, all remedies have, in regard to marching, been only more or less unsuccessful; even the device of a limber trigger has not ensured success to what may have been a good aim; a "loose hold of the gripe" still allows the sabre to come down on its flat; and the various shirkings acquired for eluding the shocks from the horse's motion give only a laboured firmness of seat, not even attaining the point of preventing an uneven riding on the two sides of the saddle, that chief cause of the giving out of cavalry horses on long marches. Still less has it given ability for the nice *perpendicular* change of the rider's weight from one seat bone to the other, which forms one of the most important of the four "aids" that give the power of controlling one's horse.

The motions of the musket-manual are, probably, based on the manner in which some perfectly formed man went through with it, but experience has shown that, for the generality of recruits, its exact execution* is simply impossible; and, so far

---

* The last external sign of a laboured, and therefore imperfect, execution is a twitching of the corners of the mouth. How few can suppress more obvious proofs!

as "carrying the musket in the *left* hand, balanced on the middle finger" is concerned, this point seems now to be about given up for a clutch with the right hand, which does anything else rather than conduce to the setting-up of the soldier.

A similar failure in exact central action, and the like eccentric results are noticeable in animals whose motions are habitually controlled by the human hand. Almost all horses bear more on one side of the bit than on the other; each has a favourite leg. Do any wear out the shoes of both sides alike?

§ 2. When we observe that the most muscular men are, often, not only less active, less supple, and less enduring than others, but also that they are not always the most powerful, it should seem probable that the force of muscular contraction often acts at a disadvantage, and is partially lost from the want of a perfectly concentric action of all parts of the bony framework on which the muscles brace themselves.

We assume then that some fault in the play of the bony skeleton is the radical cause of the soldier's deficiencies in movement, and it will be our endeavour to ascertain the nature of this fault, as also, if possible, to search out some simple and thorough means, not only for correcting its manifestations, but for restoring the animal machinery, at least for the moment, to perfect working order.

§ 3. There seems no reason to suppose that the body of man, more than the body of any other animal, was intended to have a *one-sided* action. Horses driven or ridden by one-sided men may indeed be shaped into using always the right, or always the left leg as the leading limb, or one particular side of the jaw as their favorite resisting point. Other animals which such men train may exhibit traces of similar tendencies, possibly the offspring of these animals may have them from their birth; but we see no like partiality, as to the use of his limbs or jaw, in the lion or tiger as they grasp or tear their food, nor in the squirrel, as following, with precise adaptation of his body, the direction of his eyes—he leaps from branch to branch. Some particular point reached in the respiratory action, or some casual inclination of the head would rather seem to determine any choice made by them as to which side shall commence or lead a movement.

We may add that, although with civilized man the right hand is almost always the ready hand, and the left leg the bracing leg, yet the not infrequent occurrence of the opposite peculiarities strengthens the analogical proof just given, that this is not a necessary state of things. From the Sacred History it would seem (comparing 1st Chronicles, xii—2, and Judges, xx—16), whatever may be the etymology of the Hebrew word for left-handed, that the men who could sling stones at an hair breadth, and not miss, were such as *used both hands alike.*

§ 4. On the principle that the inequality of strength in the arms was to be remedied, considerable attention was at one time paid in the British army to exercising the left arm. It might appear to be a sufficient remedy for grown men that they should exercise the neglected arm, and for children, that they should be educated in exercising both arms equally; but, in the former case, the results of so many years' previous habit having to be overcome, the remedy hardly reaches the centre of motion;* and, in the latter case, so long as children are subject to the prevailing customs, particularly those of always reading to the right,

---

* Perhaps the nearest approach to a fundamental remedy is the rule given in the French, and in the, from them translated, U. S. Cavalry Tactics to "keep the right shoulder forward," but, unless there be a good understanding that this means the upper right side of the chest, it, too, may fall under the category of covering one fault by another.

which gives an unequal exercise of direction to the eyes; and of always writing with the right hand, and to the right;* as well as to the effects of every instrument being adapted to the right hand, it should seem that some more fundamental means were necessary.

§ 5. If we may discover any movement, or series of movements, by which a man can, as it were, untwist his frame from its distortion, and then hold it steadily for a time in proper working position, we shall give to every muscle the opportunity for perfect action, and this, again and again repeated, may restore the weaker one of every pair of muscles to an equality with its fellow.

We offer the following attempt to discover and explain such movements for what it may be worth:

The action of the Spine must in the first place be clearly traced out.

§ 6. The first step in pursuance of this design must plainly be to trace out the working of the skeleton in that animal which accomplishes locomotion by the simplest means.

§ 7. The general figure of motion in man seems obvious enough (although to describe it minutely

---

*Even did we write with the left hand, it is evident that only when *writing to the left* the action would be properly changed.

be no easy task)—but the positions of every separate point of this figure depend on the movement of some more inward and more central point. Thus the hip-bone which carries the thigh-bone socket is a more inward point than the thigh-bone and is a controlling basis for its motion, as also for that of the leg and foot. So is the motion of the shoulder-blade, which carries the arm-bone socket, a controlling basis for the arm, the forearm and the hand.

Further, both the hip-bones and shoulder-blades depend for a basis of movement upon the spine and indeed finally upon the head, this last being so situated in respect to the spine that the relative positions of the articulating surfaces in their common joint, at the summit of the neck, determine the shape assumed by the spine, and consequently the angles at which the forces acting from the spine press down their points d'appui against the ground, and thus give the direction of the ensuing movement.

§ 8. The *hip-bones* are soldered into one peice with the lower end of the back-bone, and therefore depend directly upon it for every movement. The *shoulder-blades*, on the other hand, are connected with both the back-bone and neck (the two together are called the spine), not solidly but by the intervention of muscles; so that, although the arms must, in all completed motions, finally conform to the movements of the back-bone, they have yet a certain liberty of action, superior to that of the legs, and which is determined rather by their connections with the head and neck, than by their union with other parts. The arms are therefore more particularly *head-limbs*,* while the legs may be called *body limbs*.

§ 9. We, therefore, assume that, so far as the skeleton is concerned, *motion originates in the spine, is directed by the head, and is only followed up by the limbs*.

§ 10. The researches of comparative anatomists have demonstrated, that the closest analogy exists between relative parts of all the animals classed as vertebratæ (viz., such as have a spine composed of bony blocks, or vertebræ, joined by ligaments into one flexible rod). For example, that the fore-legs and fore-feet (heads-limbs of quadrupeds) answer to the arms and hands of a man, to the wings of birds, and to the pectoral fins of fishes.

---

* Comparative anatomists have shown the propriety of the name "Head-limbs" by tracing analogues of the shoulder-blades in some of the lower animals (as fishes &c.,) in actual and close attachment to the head.

They have shown that where, for lower animals, certain parts of the higher ones disappear, representative pieces may yet often be found, as for example, pieces of bone, apparently useless, but answering in position to the hip-bones of higher animals, on the bodies of some serpents. They have also made it very probable that *legs and arms are but metamorphosed ribs.*

§ 11. From this close analogy of structure, we may infer a close analogy as to the principles of locomotion among these animals, and we may, therefore, look to those in which the motions of the spine are the most obvious, and in which these motions most immediately produce locomotion, for a clue to its more obscure action in quadrupeds and man.

Assuming, then, that *the spine is the true basis of all movement;* that its deformity, brought about by permanent abnormal flexures, arising from an unequal counter-action between muscles intended to balance each other, is (where the frame is otherwise healthy) the true cause of distorted action; and, that, *to redress the shape of the spine,* in these cases, *will be to redress the faults of motion* in all parts dependent on the spine, we proceed, in the first place, to inquire how the spine acts.

PART I.

THE SNAKE'S MOTION CONSIDERED AS IN A SIMPLE ELASTIC ROD.

§ 12. The snake seems to afford the best opportunity for observing the play of the spine. Its locomotion is obviously produced, simply by the action of its back-bone upon its ribs, the remote end of the ribs being thrust against the ground, so as to propel the animal in the direction of the general resultant of all the forces developed.

§ 13. On studying the mode of locomotion in the snake, it is, we think, evident that this animal propels its body with a succession of undulatory curves, the flexures of which alternately replace each other in such a manner that those portions of the body which, during one act of propulsion, are either convex or concave, become, respectively, concave or convex for the next act.

§ 14. If we take a piece of twine, or, still better, of whip-lash (which, for description's sake, we will suppose held perpendicularly before the face), and, holding an end in either hand, turn these ends in contrary directions, so as to twist the intermediate portion, we shall find that this central part will readily form itself into flexures similar to those which a snake produces in its body during locomotion.

For future nomenclature, we may liken each division of the flexures, although the curvature be not in a single plane, to the letter S, and we shall then have an ess proper and an ess reversed alternately throughout the series.

If we next completely untwist the piece of whip-lash, and then, with the directions of the counter-turns exchanged, twist it again, we shall have similar results, with the only exception that where the curvature was a proper S, it will now be a reversed S, and vice versa. In other words, regarding only the lateral aspects, where there was a concavity to the right we shall have a concavity to the left, and where there was a convexity to the right we shall have a convexity to the left.

We shall call each separate section of torsion an S, half a section a C.

§ 15. Were the piece of whip-lash of perfectly uniform consistence, and were all the rectilineal elements of its cylinder drawn with exact equality of force, it may be shown, we think, that a series of somewhat one-sided cones, alternately point to point and base to base, would be the result. But by twisting it in the manner described, one set of elements is more particularly drawn, and these are the first ones to become stretched. It is along these elongated lines that the convex aspects of the turns occur, while the other parts are *compressed into concaves*.

§ 16. If for the whip-lash there be substituted a firm but elastic cylindrical rod, the two diagonal forces of rotary counter-traction, supplied by the thumbs and fingers, may be replaced by two forces of diagonal counter-*pressure*, the points of application of which will be situated at opposite edges of terminal horizontal planes at the ends of the rod.

Replacing then the piece of whip-lash by an elastic cylindrical rod, terminated at each end by plane surfaces cut perpendicularly to its length, the rod being held perpendicularly before the face; replacing also the action of the fingers by two compound forces, consisting each of a force of pressure in the perpendicular plane and a force of rotation, around the axis of the cylinder, in the horizontal, the latter drawing the substance of the cylinder with it; let these forces be applied, one on each terminal plane near the circumference, and at diagonally opposite points, and while the forces of pressure act in contrary directions, let the forces of rotation act also in opposition to each other.

§ 17. We shall call the diagonal points, at which the compound forces are applied, "*Points of Application,*" and as we *confine ourselves to rotation from front to rear,* there will be for every S *two diagonal pairs,* or four points in all. The diagonal pairs will replace each other in successive counter-actions between their points.

§ 18. We shall, for easier illustration, always consider an example in which the opposition of the left upper and right lower forces begins the successive counter-actions. In this the *left upper force* rotates from the front centre of the border of its plane, by the left, to the rear centre. The *right lower force* likewise from the front centre, but by the right to the rear. The action of the other pair of diagonal points will then, mutatis mutandis, alternate with these.

§ 19. The *theory of the twists* may then, we think, be discussed as follows:

If we first bend the rod in the simplest manner, *i.e.*, by pressure, without rotation, applied at perpendicularly, not diagonally, opposite points, so that it shall take the shape of a C, then a spring will be formed; and if, while one end of this spring is fixed, the other end be liberated so that it can pass in one line only, the force of the spring will be exerted in the direction of that line. For example, if we suppose a straight tube of smaller diameter than the depth of the concavity of the C, to be held over the upper end of the rod, while the lower end is immovably fixed, the spring will discharge itself in the line of the tube.

§ 20. Let us next apply the counter-pressures at *diagonally* opposite points. The pressure from either point forms, as before, a concave beneath the point, but this concave no longer forms one with that under the other point. On the contrary, the concaves being now on opposite sides are each met by the pressure of an intermediate point, and form an S, or two C C with their hollows in opposite directions.

The straight line joining the two points of diagonal pressure, must cross the rod in a part of its length determined by the nature of the rod, and the ratio of the pressures to each other. In this crossing all *counter-twisting* forces from the points of pressure will meet, and here more especially oppose each other.

We shall designate by the expression "*Crossline,*" this portion of the line joining the two diagonal points. This "cross-line" plays a very important part in the following theory of locomotion, and as there will be two such lines, the one produced by the antagonism of the left upper and right lower forces—the other by that of the right upper and left lower, we shall discriminate the *cross-lines* as *left-right* and *right-left* respectively. We shall also name that end of either cross-line on which an upper force draws the *upper end*, and that on which a lower force draws the *lower end*.

§ 21. It is particularly to be observed that *the curvature under any point of pressure is always concave.*

§ 22. Straightening the rod again, let the point at which the upper diagonal pressure is applied (according to our standard example the left upper) be placed at the front centre of its plane, and then moved in rotation horizontally, the pressure continuing, and let the lower end of the rod be prevented from following the movement so that it offers merely a passive resistance to it. There will then be a constantly increasing pressure exerted below the moving point by the reaction through the stretched linear elements of the rod.

The direct effects of the traction will be in a *winding line* passing from the point of application, by the front, half way around the rod, to the point where it meets the resistance from the stationary lower point of application, and which point is in effect the "upper end of the cross-line."

So far as the rotary traction from the upper point continues its course, by drawing on the upper end of the "cross-line" (here situated on the right side) it will produce a continuation of its winding line from that end, by the rear, half way around the lower part of the rod, to a point at the lower plane section, perpendicularly under and of course collateral with its (the upper left point of application) own position. We shall thus have a helical line completely around the rod from a point on the border of the upper plane section to a point perpendicularly under it on the border of the lower plane section. This line, however, is interrupted in its continuity of force by the "cross-line," which (kept in place by the passive resistance of the lower point of application) continues the winding line in just such proportion as it is moved from its place against the resistance of its other end.

§ 23. The drawing on the substance of the rod along the winding line will, by the reacting pressure, produce concavities under the varying positions of the rotating point, and the line of these concaves, as also of their corresponding convexes, will, of course, likewise be helical.

So far, also, as the end of the "cross-line" is drawn into rotation it will produce concaves *under* its course in the continuation of the winding line.

Were the diagonally opposite lower "point of application" in active working, it would produce another winding line in the opposite direction, under similar conditions, viz : from the right side of the lower plane, by the front, to the lower end of the cross-line, and thence (this cross-line end being moved) completing the circuit, by the rear, to the right side of the upper plane section.

§ 24. Two other such "winding lines" may be developed from the other two diagonal "points of application." At present we only consider the diagonal points as acting by pairs, alternately. The consideration of the two pairs, acting simultaneously, will be taken up further on.

§ 25. It will be observed then, that each of these winding lines consists of two sections ; the *primary* one, which passes from the point of applicacation to its end of the "cross-line;" and a *secondary* one, formed by the traction from the cross-line end, which passes to a point situated in a perpendicular line from the originating point of application.

The twisting of a cross-line is caused by the

meeting of the *primary* sections of the two diagonal winding lines. It will be observed that the secondary sections terminate each respectively at what will be the point of application of the alternate opposite line.*

§ 26. We shall designate each of the four winding lines by naming its point of departure which, since their courses are entirely distinct, will fully distinguish it thus: *Upper left winding line—lower right winding line—Upper right winding line—lower left winding line.*

§ 27. As regards the concaves produced by the revolution of a single point of application against the simple *passive* resistance of its diagonal point, those concaves more directly under the active point will be the deepest and shortest, and *in each C there will be two gradations of curvature, of which that most remote from the active point will be the longest and least sharply inflexed.* This last point is of considerable moment.

§ 28. So soon as either one of a pair of diagonal forces, drawing on its winding line, meets a certain amount of resistance through the cross-line, whether from the active working, or the passive resistance of the other point, its traction will be-

---

* To complete this section see note, § 39, and particularly § 201.

come more or less completely absorbed in the cross-line, where it has to meet the contrary turn of the opposite cross-line end.

Each point of application will move its end of the cross-line in a direction contrary to its own course. Thus, *the left upper point of application will move its (the right upper end) of the cross-line to the left, by the front,* and *the right lower point will move its end (the left lower) to the right, also by the front.* A new S cannot develop in the cross-line, but the influence of such an S will spread out from it into the C C of the original curve.

This new S, arising in the centre of the old one, from the reaction of the cross-line, will be twisted in precisely contrary directions to the latter, and so far as its influence develops, will tend to reverse the original curvatures. In this way *each upper end of a cross-line will, as it were, turn up into the C above it,* and *each lower end down into the C below it—reversing the curvatures from their positions and carrying a similar general effect throughout the C.*

This principle, that every set of torsions will, when carried to a certain point, tend to reverse themselves, lies at the foundation of the theory of locomotion.

§ 29. The new curvatures advance, not as the old ones, from the extremities to the centre, but from the centre to the extremities, and, if the original curvatures could be discharged, and the extremities moved across each other, we should have the first S replaced by another of reversed flexures, and we should have the discharge of the original curve of the rod so constrained by the nascent one that its direction might be made entirely perpendicular in both C C; in a similar manner but much more accurately than the discharge of the simple C curve mentioned in § 19, was constrained by the tube held over it.

The power of the nascent new shape will be largely exhausted in thus constraining the direction of the spring from the old, but a portion would remain to join the alternating points of application in impressing the similarly shaped new curve upon the rod.

§ 30. If we examine more closely the lines of direct and of reacting traction we think that the manner in which a change of curvatures would take place, if appropriate free articulations in the course of the rod allowed the discharge of the first curve, may be explained as follows:

The contrary curvature arising in each C is constraining the direction of the old curvature throughout, and incipiently altering it at the cross line end. Now, if we conceive of a half revolution with a ball and socket joint at the junction of the two C C, and at the centre of each, we can suppose the escape of the old curve, and that the alternate diagonal forces, coming immediately into play, accept the central parts of the new C C from the cross-line ends, and, forming the remote parts from the extremities, join in reciprocal counter-action by new cross-lines.

The reversal of the curvatures would be accompanied by a spring, in each direction from the cross-line as a point d'appui.

§ 31. The above gives a spring in two directions, and no advance is made. To accomplish progressive or retrogressive locomotion the spring must have place in only one direction with a point of first appui at one of the extremities of the rod, in order to move in the direction of the other. The cross-line gives the requisite appui for the C remote from the fixed appui (the ground or other basis), and the reaction between this and the cross-line furnishes the spring for the nearer one, and also sustains the spring of the first.

§ 32. After the upper point of application has formed the general curvatures of its winding line, its further action will displace its end of the cross-line toward its own side. This displacement of the upper end of the cross-line will check the similar movement of the lower end to the opposite side, when the diagonal lower point of application acts. Consequently when the "lower point of application," forming its "secondary section" first, has prepared the upper C for discharge, by introducing into it a counter-traction and twist, and, next, comes to form its own "primary section," by moving the lower end of the "cross-line," it will find this end immoveably fixed, and the forces generated by its rotation will thus be brought to bear against the appui of lower end of the rod on the ground* (not against the cross-line) and be kept against the ground until the last moment of the general discharge, when only, the lower end of the cross-line may be moved.

Thus, the upper C being discharged, the lower C will immediately follow, the latter acting, not against the cross-line, but in harmony with the upper C against the upper terminal plane.

§ 33. The point of appui on the ground for the lower end will be on the side to which the whole cross-line is maintained during the action, *i. e.*, the

---

* If the lower cross-line end be fixed to the left, this appui will be on the left side, and vice versa.

side collateral with the upper point of application.

§ 34. Were the rod laid horizontally on the ground with the lower end at the rear, and the ground appui of the now rear end provided for by some projection, the now anterior C might at its convex side be provided with an independent appui on the ground, which should aid, not in projecting it forward, but in forming the counter-actions which give the spring. *This additional appui*, by collecting force from the anterior C, would raise it from the ground, and not being on the side first thrust forward by the changing curvature, *would leave the ground after the rear point of appui.*

§ 35. We have considered only one S curve, but, if there were to be a succession of them, the action generated from one "point of application" might be transferred from one S to the other until all were thrown into form. In this case the junctions of every two C C (of the S S, marked off from the top downward), even when they belonged to different S S, might all be considered as contrary "cross-lines," in reference to each other, but we shall find it more consonant with facility of explanation to suppose fresh "points of application" at the terminal planes of junction of all S S, and "cross-lines at the junction of their C C."

PART II.

APPLICATION TO THE MOVEMENTS OF THE SNAKE.

§ 36. The actions and reactions which we have described for the rod are, we conceive, those which take place in the body of the snake, and which produce its locomotion.

The two pairs of antagonistic forces, as then brought into play, represent and are the resultants of the general muscular action of the animal, added to the elasticity of its ribs.

The various articulations of the spine, head and ribs allow the discharge of the different alternating twists.

§ 37. The head of the snake is coupled to the back-bone by a perfect ball and socket joint. Of the two parts of this joint, the ball is carried by the head, and the socket by the head-end of the back-bone.

In descriptions of the snake's action we must change the terms *upper* and *lower* used in relation to the rod, to *anterior* and *posterior*.

The head-joint then being a ball and socket, the equivalents of the "points of application" and the results of the "winding lines" in the rod, would, as represented at the head-joint, in its actual or relative movements, be as follows. We take the left upper and right lower counter-actions.

(*a*) Accompanying the formation of the *left upper winding line, the left side of the head-ball,* acting as a point of application, *revolves outward and backward.* This extends to its cross-line end, and then forms the secondary section of the anterior winding line.

(*a'*) Accompanying the movement of the upper end of the cross-line, the *pressure of the head-ball turns in its course, passing forward and to the left,* a movement which, when constrained by the action of the right posterior line, is made directly forward. This moves the cross-line ends and draws tight the primary section of the upper winding line.

(*b*) The tractions of the right lower winding line cannot work at the head, and those of the left upper line be maintained, unless by moving the socket. *The drawing of the right lower winding line would rotate the right side of the socket outward and backward.* This extends to its cross-line end, and then forms the secondary section of the lower winding line. Its action at the anterior end of the socket resembles that of the condyle at its posterior end.

(*b'*) The action of the lower end of the cross-line

will give to the socket a forward movement with a turn to the right, but when constrained by the action of the other winding line directly forward, a' and b' combined tend to liberate the head-joint, but, as will be noticed further on, the locomotive gathering requires that a considerable part of b' precede b.

§ 38. The outward motion of the right side of the socket, if transformed into motion of the left side of the head-ball, would double it, and the same for the motion of the head-ball in relation to the socket motion.

§ 39. We think, then, that we may, for our purposes, describe the movements, actual and relative, in the head-joint, as follows, for the left-right opposition of forces.

(*a*) *The head-ball* performs a rotary movement to the left and backward, pressing downward on the left lateral hemisphere of the socket. This pressure passes forward.

(*b*) The right-half hemisphere of *the socket* performs a rotary movement to the right and backward, at the same time *drawing itself away from the corresponding part of the head-ball, and consequently raising the left side of the socket under* the active side of the head-ball, and increasing the pressure exerted by it.*

The socket also moves forward.

§ 40. It will be observed that the outward movements, both of head-ball and socket, are in opposition to the central forward movement of the head-ball, and the force of each movement being retained in the body by the setting of the muscles, there will be a straining point between the head-ball and socket, about the centre of the joint.

If the advance of the socket, which finally relieves this strain, were made in anticipation of the outward rotation of the socket, the joint would not be freed until this rotation were accomplished.

§ 41. These different positions of the head-ball answering to those of the upper, now front points of application, act in conjunction with a greater or less number of subordinate front " points of application" at the junction of the different S S formed in the body of the snake. As mentioned in § 35, these points are similar to the cross-line points, but, it seems to us, that it is only the section planes

---

* This separation is caused by the posterior winding line, and this line participates, throughout its course, in having its *elements of pressure reflected* to the opposite side, as are those of the separating socket, in fact, to the convex sides.

across every second change of C which *primarily* act in connection with the originating movement, or winding line formed by the *leading* point of application. The intermediate planes do not become planes of pressure until the movement of the *subsidiary* point of application [*i. e.*, not the leading point] affects the cross-lines. For this and for other reasons, we find it more convenient to treat of the planes at the commencement and ending of S S as containing points of application.

§ 42. The winding lines from the head having been formed, and a succession of helical turns shaped against the *passive* resistance of the successive posterior points of application, then *active* counter-action begins from the rearmost of these latter points, and its effects are transferred forward from point to point, *actually establishing* first in the *anterior* cross-line, and then in each succeeding cross-line the latent reversal of their respective S S.

The whole force is thus centred on the head, and when this, by its actual or relative movement, releases the front point of application, the winding line from the rear point cutting, as it were, through the body of the snake, *allows the development of the spring*, and becomes, on the opposite aspect of each C, the new alternating anterior winding line.

In this case, where we have begun with the left anterior and right posterior, the latter becomes the right anterior line leading, and the left posterior will develop on it.

§ 43. Although the general action is the same, and, on the theory of each interior point of application being in a cross-line, we might consider each C from the front as replacing the one in rear, and each from the rear as replacing the one in front, yet we shall, for reasons which will appear when the locomotion of the higher animals is taken up, first consider the action of each as simple and unconnected with others.

In general action, the head is steady, and the spine moves from or against it at the socket; but, since the forces are gathered against the head-ball as a focus—since the ultimate result is as if the head gave a final covering twist, and since it seems to facilitate explanation—we shall *suppose the head to move.*

§ 44. The *spine* or back-bone of the snake, which represents the simple elastic rod of the preceding discussion, is made up of a large number of little blocks of bone called vertebræ. These are jointed to each other by means of a convex surface on the rear of one vertebra, fitting into a concave surface in the front of the next. Thus the utmost freedom of motion is allowed, and the numerous powerful muscles make of the spine a rod of almost perfect elasticity, and capable of all the necessary adjustments.

The *ribs*, by which the snake must evidently take its final appui for all motion, are set by pairs—one rib on either side of every vertebra, so that the courses of their articulations form parallel lines, from head to tail, on each side of the spine. These articulations are formed each by a socket of two slight concavities on the upper end of the rib, moving on a protuberance from the vertebra which carries corresponding convexities.*

Thus set on, the ribs support the spine like so many curved springs bowing outward.

At their ground ends each one of a pair of ribs is connected with its fellow by a *ligamentous band*, and these bands offer the medium by which, in transverse continuation of the lower ends of the ribs, the animal takes hold of the ground.

§ 45. If we call the position of the *ribs* in their sockets, as the snake lies extended, their *normal position*, and assume, for the moment, that the rib

---

* This differs from the analogous articulations in the higher animals, where the rib carries the ball, and the sockets are *between* two vertebræ.

does not move in its articulation, then, when, by the formation of torsion curvatures in the spine, the facings of the protuberances on the vertebræ are changed, viz., to the front, by coming on the anterior portion of a convex, or the posterior portion of a concave, and to the rear by coming on the posterior portion of a convex, or the anterior of a concave—it is obvious, that the facings of the ribs will be changed correspondingly. The concave inner surface of each rib will be in the snake turned toward the rear when the ball of its articulation is turned to the front, and to the front when the ball is turned to the rear.

We shall always speak of a rib as "facing forward or backward," with reference to its concave surface. Thus the ribs on the anterior half of convexities, and posterior half of concavities, face backward, the ribs on the remaining halves forward.

§ 46. If next, the ribs in any facing be pressed against the ground, so that their ground ends are firmly fixed; then an altered facing of their articulations, such as would be caused by the commencement of a change to the opposite curvatures in the spine, will introduce a twist into the C shape of each rib, thus changing it into a twisted S; and when this twist is discharged at the articulation, in the manner we are about to describe, one of the turns will give force, forward or backward, to the ensuing spring, and, as in the case of the spine, or rod, the other turn will control the direction of the spring.

§ 47. We may consider the cylinder formed by the spine, the ribs, and the ligamentous connections between the rib ends, as a *compound spine*, in which the idea that the elastic rod of our previous discussion should be able to release its twists of one form, so as to accept those of the replacing form, is carried out.

If this be allowed, we see that the *fundamental action* of every portion of the machinery for *locomotion* is the action of *the double twist*, viz., a turn in one direction met by a turn in the contrary direction, and under the rule that one of these turns being liberated it is guided as to the direction of its discharge by the constraining influence of the other.

For example, a rib faced with its concave forward, by reason of the contour of the spine, under its articulation, and, becoming twisted by a turn in the contrary direction, will, finally, with the reversing action of the cross-line of its spinal S, spring at the moment when the shape of the spine is changed. *It will then be the primary turn which discharges itself against the spinal articulation,* and this gives the locomotive force, the other turn merely guiding the direction.

The primary turn would be in the reverse direction for a different succession of twists, viz., such as would have place from beginning with a rear point of application, but, if we be not mistaken, the ribs of the common snake are, normally, so inclined as to bring the concave surface to the front, when they are not in action, a circumstance which would indicate that the final slip at the articulations is always forward. We expect to show how locomotion backward may be produced with an anterior point of application leading, and it is, we suppose, for this reason, viz.: retrogression being, in vertebrate animals, derived from progressive action, that motion backward is somewhat awkward in comparison with the motion forward.

There is one species of snake, the *amphisbæna,* which, it is said, moves with equal facility in either direction. Whether in these the ribs are so set on that they may discharge by a slip backward, as well as forward, and so, readily, interchange the leading points of application, we are not able to say.

Of the two articulating surfaces on the rib protuberance (§ 44), which constitute the ball portion of the rib and spine-joint in the snake, we should imagine that the rearmost one receives the pressure when the concavity of the rib faces forward, the anterior one when it faces backward.

§ 48. Returning to the action of the head-ball, and of its socket (which latter is carried by the first vertebra of the spine), we will endeavor to carry out the principles stated, to a connection with the ribs, &c., when concentrating the spring for an act of progressive locomotion.

(a) (§§ 39, 37). The head-ball rotates from the left, by the rear, and toward the right, making pressure in the left hemisphere of its socket, under which a concave forms and its continued action forms [against the *passive* resistance of the right rear point of application, *i. e.*, one of the rear rib articulations on the right side], the "left anterior winding line" (§ 25) in its *secondary* section, *i. e.*, in the posterior C of the S.

(a') The continued rotation of the head-ball, against the passive resistance of the right rear point of application, after it has formed its secondary section brings around the upper end of its cross-line, to the left and front, forming fully the *primary* section of the left anterior winding line. The head-ball passing further around comes to a check, so as to press against a point toward the front of the socket, and somewhat to the left of its front centre.

(b') (Which, in a part of its development precedes b, § 37). The now commencing *active* working of the right rear point of application does not at first form the secondary section of its winding line, which would form in the anterior C of the S, that being prevented by the full formation of the anterior line which has displaced the upper end of the cross-line, and thus checks the movement of the lower end, so, indeed, that this end cannot fully draw until the discharge of the anterior line allows it to come again into traction. The action, then first, forms part of the primary section of the rear line, and introduces the change of curve, from the cross-line end, first *into the rear C giving the counter-turn to the ribs along its convex.* Its effect in the head-joint is to move the socket forward, but at the same time with a turn to the right, which brings the left anterior line point of pressure back to the centre from its inclination to the left mentioned under (*a*).

(*b*) The continuation of the active working of the right rear point of application next, causes its secondary section, which is the nascent reverse curve to begin in the anterior C, spreading from the *upper* end of the cross-line, as its formation forces this end backward from the forward position into which it has been drawn by the formation of the primary section of the anterior line. This nascent curve gives the second turn to the ribs along the convex of this C, as the partial development of *b'*, which is the nascent reverse curve for the posterior C, did to the ribs on its convex.

Finally the commencement of the alternate (here the right *anterior winding line by the movement of the right anterior point of application, beginning with the development of its secondary section (a) will first discharge the rear C*, and so on; the old right rear winding line becoming the new anterior winding line, by cutting through the articulations. But a closer examination of the action of the ribs, &c., will show more clearly how each point discharges its spring.

It is evident that, if the head were gradually shifted to the right so as to bring the posterior point of application to the same relative position it would take by active movement, all these effects could be brought about by the movement of the left head condyle, against the *passive* resistance of the rear point, continued throughout.

§ 49. As has been already mentioned, the number of S S formed in the elastic rod, as well as their relative proportions, would be dependent on conditions involved in its structure; so in the snake's body, where these conditions must depend very much on the will of the animal, the number of S S and their proportions may probably be regulated at its pleasure. In man and quadrupeds, however, they are fixed, in both respects by the form of the mechanism.

§ 50. There is one other set of actions which might be here discussed, viz., those arising from both pairs of diagonal forces acting simultaneously, but, as it seems to us doubtful whether the snake, having no unyielding breast-bone, be capable of using them, and since they may be as well and more conveniently taken up, when speaking of the higher animals, we defer them for the present, excepting, so far as they are spoken of in § 71 and onward.

§ 51. In the different facings of the protuberances of the spine, with which the ribs articulate, the ribs will be thrown more or less upon the outer or the inner edge of their ground ends, and, being bent against the ground on one edge, the introduction of a counter-turn, though this be only potential and latent as regards the spine, will yet bring the bearing of the ribs on the contrary edge, to that on which they at first rested.

We believe that two bevels—an inner and an outer—are found on the ground ends of the snake's ribs; but, however this may be, *the change from outer to inner side (or vice versa) of the feet which furnish the appuis on the ground in the higher animals is a marked feature of their locomotion*, and we may observe a no less marked distinction in the successive application of the two sides of the palm of the hand in man, when this member is perfectly used.

We shall often employ the terms *innner* and *outer bearings* (or *bevels* when speaking of the edges of the ground ends of the snake's ribs, and also of the two sides of the feet in quadrupeds, and of the feet and hands in man.

§ 52. It will be noticed that, in the snake, the helices of the spine appear from the extreme pliableness of the ribs, to be more or less flattened, so that the body seems to move chiefly by curvatures in the horizontal plane. The spine, however, retains its *helical* curvature, and there is real action in the perpendicular plane.

§ 53. We will then suppose that the ground end of each rib is terminated by *two bevels,* so cut that *on* one of them—*the outer—the rib shall rest when faced to the front* (§ 45) and before receiving any secondary turn ; *on* the other of them—*the inner —when faced to the rear* and under the same condition. Then the *ribs* which are in appui according to § 48 *(a)* will, before they have received a second turn, rest as follows : those *on the anterior part of a convex* (being faced to the rear) *on their inner bevels* —those *on the posterior part of a convex* (being faced to the front), *on their outer bevels.* Should the ribs of concaves be put in appui the order would be reversed.

§ 54. The forces act, primarily, along the convexes, and the appuis are *normally* on the ribs articulated along the two convexes of each S. Should the ribs of a concave take the place of those of a convex as appuis, the forces would act on them only secondarily, *i. e.*, as a sequence of the action along the convexes.

Supposing the left anterior point of application in action : As the winding line is formed and the anterior end of the cross-line is drawn around to the right, the ribs on the right of the first C and on the left of the second (here being the two convexes of the S), will be in appui, and those on the concaves will be raised. The ribs along the posterior parts of the convexes being already on their outer bevels, the further traction from the left anterior point of application, when it begins to move the front cross-line end will increase this bearing on the outer bevels.

§ 55. The effect of continued helical traction of the left anterior winding line throughout the convexes, that is to say, in both the posterior and anterior convex, is finally to turn *all* the ribs articulated along them *outward*. Even those in the front portions (faced to the rear) will be thus affected, so soon as, with the straightening of the line, the upper cross-line end permanently moves, and they come under its direct influence.

The posterior part of the *secondary* section of the line (§ 23) is first affected; then the upper end of the cross-line, being somewhat moved, affects the posterior part of the *primary* section, and so on until the secondary section, having reached its limit, the upper cross-line end is more absolutely subject to the traction.

The effect of the posterior winding line on the same articulations, along the convexes, is to turn them inward.

The reciprocal cross-cutting of these lines in the convexity articulations is the means of their discharge with the spring which they have gathered, so soon as it is liberated.

§ 56. The double twisting of the convexity ribs may receive and retain the elements of the discharge, as produced in them by the diagonal winding lines, before the curvatures of the spine are at all affected.

As was mentioned in § 32, the displacement of the cross-line by the anterior winding line—in this case to the left—brings on a reaction to the posterior line from the head, and the rear cross-line end will not draw until a final exertion of the rear "point of application," or a redoubled working of the anterior point (§ 38) restores it to its place.

§ 57. We should suppose the discharge to take place as follows:

The left anterior winding line, having established the right anterior, and left posterior, convexes. The posterior line works—with reaction at the head—from the rear of each convex; and the first effect of this working is to double-twist the ribs on the rear posterior halves of the convexes, and change their outer to an inner bearing (§ 51).

This being done, the final movement draws the posterior end of the cross-line and anterior termination of the right posterior winding line [which results in the outward turn to the rear of the right hemisphere of the head socket (§ *b*, § 39, etc.),] so into place, that the winding lines cut each other through the *anterior* halves of the convexes, also changing the inner bearing of their ribs to an outer one. The discharge is thus virtually completed, and the body in position for the alternate gathering, if the alternate anterior point of application come at once into play.

§ 58. The posterior halves of the convexes, being virtually discharged by the complete drawing—although not the cutting—of the lines, are thus actually discharged; the *posterior*, when the posterior cross-line end receives the full traction of the posterior winding line; the *anterior*, when the alternate anterior point of application begins its working. Or, perhaps, rather the whole discharge is only virtual, until this last action occurs, when the actual discharge and spring rapidly take place, beginning with the rearmost appui, in the succession mentioned.

It will be remarked that the course of *the cutting line* coming from the opposite side, *for the ribs on the posterior halves*, the line first takes effect in the trunk, and thence *ascends to the articulations*, whereas, coming for the anterior halves from their own sides, it *descends at the posterior cross-line, and at the head-ending of the posterior " winding line," from the articulations, into the trunk.*

In these movements the secondary section working of the anterior winding line, in the posterior C, represents the turning (a) of the head condyle. Its primary section working, including the action of the anterior cross-line end, represents (a') the passage of the condyle pressure across the head-joint to its front. The secondary section working of the posterior winding line in the anterior C, represents the turning (b) of the socket by its outer edge. Its primary section working in the posterior C, which partially anticipates the secondary, because the posterior cross-line end is checked in its movement, represents (b') the passage of the socket action across the head-joint to the front.

On the sides, also, the crossing of these lines could be projected as ball and socket surfaces, the anterior line, then, tracing the sockets, as it produces the concaves of the alternate curve, and the posterior line the balls, as it produces the alternate convexes.

The figure of 8 *shape, as developed in the ball and socket movement,* one half by the ball and one half by the socket, or, by the cutting of two contrary helices, *seems to lie at the foundation of all locomotion,* and perhaps of all kinds of motion, including that of the final atoms, pressure and rotation being its elements. This connection of an 8 shape with locomotion has, we believe, been noticed by several authors.

§ 59. It will be noticed, further, that (§ 27) *the posterior part of each C,* as formed by the leading point of application (here the left anterior), *is the longest,* being furthest from the originating point, and consequently carries the greater number of ribs. The anterior ribs of each convex, then, discharge their inward turn causing a spring backward (§ 47), against the action of the numerically preponderating ribs of the posterior parts, which discharge their *outward* turn, causing a spring forward. *The anterior ribs of each convex will then take the role of lifting the section while the posterior ribs drive it forward.*

§ 60. So soon as the continued advancement of the *potential* change of curvature extends its effects *actually* to the spine, a discharge is prepared in which the posterior C is so far discharged and lifted as to crowd forward the effect of propulsion on to the anterior C, and their united discharge thrusts the body forward by a movement in which the *rear ribs of an S* really precede the front ones in leaving the ground.*

The changing of the curvature of the convexes from one side to the other has, owing to the convex being the part which primarily follows the lines of traction, this characteristic, that *the convex passes over, the concave under*, in making the change.

§ 61. The *ribs on the concaves do not change their* facing *until the actual change in the shape of the spine* brings this about by changing the direction of their articulating heads.

§ 62. The above stated conditions, when each section of ribs has been twisted by the turns last introduced into a position for discharging the primary ones, at once divide the ribs on any convex, or concave (should such be in appui), into two sets,

---

* The same crowding forward of the action would occur were there a number of S S.

viz., those about to discharge in the direction for progression, in the one case, or retrogression in the other, which we shall term PROPELLERS, and those about to discharge in the opposing direction, which we shall term BEARERS, since they raise the body sufficiently to clear the ground.

If the torsions by both anterior and posterior points of application work together so that neither end of the cross-line shall (as in § 32) displace the other, the spring will take place against the cross-line as a centre of appui, and the result will be a *perpendicular locomotion*, in which case there will be an equal division between the bearers and propellers in each C.

§ 63. If the rear point of application were made the leading point, the reverse of the preceding action would take place, and in a snake whose structure allowed this mode of action, retrogression would follow. In such a case the ribs discharging to the front would be "*bearers*," but as this mode of action will not be considered normal we confine that term to *the anterior ribs in each C*.

§ 64. As to the order in which the *alternating* ribs come to the ground, this will depend upon the degree of gathering attained while "en air." The propellers of the rear C, as it were, running over their bearers and receiving the proper lift from them, quitted the ground first; then these bearers; next, the ribs of the front C, in the same order. Now, as, in this way, the rear C alters its curvature before the front C, and this alteration commences at its rear, the alternate propellers of the rear C will, the first, be ready to take the ground, then its bearers; and in the same order for the front C — that is, the new appuis will ground in the same order that the old ones lifted. To accomplish this, we suppose the gathering "en air" must have proceeded as far as the drawing of the right (alternate) point of application upon the anterior end of the alternate cross-line, and the passage of the head-ball pressure to the front. (§ 48.)

But, there are two conditions which may cause the anterior C to reach the ground in advance of the rear C. 1st. If the bearers of that C have not sufficiently raised it, in which case, even if the propellers of the front C should land before the bearers, the anterior line-gathering would be made on the ground, and in a somewhat awkward manner; 2d. If the spring approach the nature of a jump, and the gathering "en air" proceed as far as the commencement of the active working from the alternate posterior (here left) point of application, then the bearers of the front C in the new curve — the head end of each S being the most weighted and the sustaining force having exhausted itself — would be the first to reach the ground.

§ 65. It may here be remarked that, to ensure smooth action and to maintain the head in unswerving steadiness, it would seem requisite that two or three vertebræ immediately behind the head should be free from the ground, and thus act as a *neck* or *adjusting connection between the head and* the median point between the two first ribs which, as their articulations, alternately come to the centre line, must otherwise give it a certain lateral motion.

§ 66. The *Scutæ,* or ligamentous connections between the ground ends of each pair of ribs (before alluded to) being flexible, accommodate themselves to the position of the bevels. They act, we should suppose, as follows : When the propellers come to the ground, on their outer bevels, the posterior edges of their sides of the scutæ press against it, when they take appui on their inner bevels this edge is pressed *downward* against the ground to secure a firm resistance in appui of the gathering.*

The twisting of the scutæ also may have some influence in adjusting the bevels of the ribs "en air."

§ 67. We will now give a short resumé of the foregoing theory of the progressive locomotion of the snake.

There are for each diagonal spring *two virtual actions at the head-joint.*

First. That produced by the moving pressure of the *head-ball* and consisting (a) of the passage of the pressure around the edge of the socket on the side of the active anterior " point of application." (In our example the left). (a'.) Of its passage from the rear, across the middle of the socket, and (unless the opposing action have begun) to its own side of the front centre.

Second. That produced by the moving and retraction (as to the moving † side), of *the socket*, under the side of the head-ball, opposite to that just

---

*That is, only the posterior edge acts (speaking of the propellers). The backward bearing of the spring in *perfect* motion is included entirely within the rib or limb of appui, that element of the force not coming to the ground—were it not for this no locomotion could be obtained on ice where the weight pressing forward secures the gathering, but would not secure the spring.

† *I. e.,* side which leads the movement.

mentioned, and consisting (*b'*) of the bringing back of the point of pressure to the front centre, and that moving forward of the socket which prepares the liberation of the strain forward (*b*). The further movement of the socket by rotation outward in the opposite direction to that of the head-ball. This is more particularly accompanied by retraction of the moving side of the socket, and by raising the other side it redoubles the pressure of the head-ball.

We have called these actions "virtual," because they may, during a considerable part of their movement, only *represent* the tractions of the twisted lines of the spine.

§ 68. *The actions in the body of the snake which accompany these actions of the head joint*, are

Corresponding with (a), those of the secondary sections of the anterior winding line (§ 25), causing the preliminary part of the formation of a greater or less number of S S curves, of which the posterior C C are longer than the anterior, and also the posterior part of each C longer than its anterior part. In this way the posterior curvature of each C carries more ribs than the anterior, and thus, in each C the "propellers" preponderate over the "bearers." The "propellers" rest on their outer, the "bearers" on their inner bevels.

Corresponding with (a'), the drawing on the anterior ends of the cross lines, which takes effect chiefly in the *primary section of the anterior winding line* (§ 25). This produces an increased bearing of the propellers on their outer bevels and, to some extent a turning outward of the bearers.

Corresponding with (b'), the check which would be given to the rear points of the cross-lines, in following the anterior points is, by the strong working in the *primary section of the posterior winding line*—the rear points being prevented from moving (§ 32)—received at the head. It brings the propellers on to their inner bevels, double twisting them—first those of the rear C C, by action properly belonging to b, then those of the anterior C C, by action of the secondary section; and this through the S S in succession, from rear to front. So far as it influences the bearers it turns them outward.

Corresponding with (b), the full formation of the *primary section of the posterior winding line*, as the rear ends of the cross-lines come into place; the turning outward of the bearers and complete bringing to the discharge point of the propellers by turning them inward.

Finally, corresponding with action (*a*) of the alternating side of the head-ball, the liberation of these gatherings and the instantaneous formation of the corresponding movements under (*a*).

Remarks.—The discharge *actually* alters the shape of the spine, giving the final adjustment to the articulations of the concaves for taking the ground.

If the motion be of a rapid, powerful character, the hinder S S will not only be first brought to the point of discharge, but will first leave the ground, and be the first to come down; the whole body, however close to the ground, being in air at the same moment; but if the movement be more sluggish, the front S S must the first leave and the first come to the ground.

The point that (*a*) (*and even a'*) *form en air, when the locomotion is once begun, is a most important one.*

It will often be most convenient to describe the whole course of an action by referring it to the head-joint movements, without going through the details of the spinal workings.

§ 69. From the position of readiness to discharge, where (*b*) is carried out, *three* different results may follow, according to the further movements which may be given to the head-ball and socket-joint.

First result—progression, as above described.

§ 70. *Second result—retrogression*, if, instead of at once beginning the alternate movement of the right anterior point of application, the discharge be refused. Then any further movement of the left side of the head-ball being prevented, and the attempted rotation of the right side of the socket outward being continued, this latter movement will be reduplicated in its effects on the left side of the socket, and, being inoperative in reducing the position of the ball on the same side, will, by resiliency, take effect in *raising* the opposite (right) side of the socket and carrying it forward. In the same way, the ball motion of the left side being reduplicated, the ball will be pressed down and carried backward on the right side. We shall thus have the head-joint in full position of the *alternate* gathering, while the ribs, &c., still retain the old one.

The effect of these actions on the points of gravity will be to throw, by the first, the appui, off the convexes, upon the concaves of the first

C C; and, by the second, upon the concaves of the rear convexes, in each S.

As now the movement of the socket discharges the first, *the front* C C *will be discharged the first*, then the rear C C, in each S separately; and, the appuis now being on ribs faced in the exactly opposite directions to those on the convexes, the movement will be backward instead of forward.

The ribs of the concaves, as thus used in retrogression, will leave and take the ground from the opposite bevels to those from which they would have left and taken it had they been on a convex in progression.

The convexes, being still the parts primarily affected by the tractions (§ 54), and the appuis, now only changing their bearings with the actual change of the spinal curves (§ 61), the springs of the two C C in each S will be more synchronous in retrogression than in progression.

§ 71. *Third result.* The effect of an attempted *repetition* of any one of the four movements of the ball and socket on the same track, that is not allowing any transformation of one movement into another, will be to pass a part of the gathering to the opposite side, or, rather, perhaps, to give back part of the gathering belonging to that point of application which may be supposed to have just discharged.

By thus forcing in succession *all* the movements, after a full diagonal gathering, we obtain — as we would by the *synchronous* action of both sets of diagonal forces—that *superimposition of torsions* to which reference was made in § 43, and in regard to which doubts were there expressed as to the capability of the snake to use this double gathering in *continuous* locomotion.

By "*superimposition of twists*" we would designate the result of both the pairs of diagonal forces being in action simultaneously. The *lateral shapes* of both curves will then be suppressed, although the corresponding forces will still lie latent in the spine. On the other hand, the shapes in the perpendicular plane will remain, since they do not act against each other.

When these curvatures in the perpendicular plane are fully gathered, the junction of the two C C of each S, where the convexes formed by the two diagonal sets of forces cut each other, will be depressed, as will likewise the anterior and posterior seats of the " points of application," where the courses of these points cut each other. Between these the intermediate portion of each C C will rise and there will be formed for each S two arches in the perpendicular plane, one for each C. *The upper lines of the arch repesents the opposed convexities which have met in their passage* OVER (§ 60). *The lower lines the corresponding concaves.*

§ 72. We will now follow out the details of *forcing*, or attempting to repeat, the various movements of the head-ball and socket—assuming the usual order of the actions, and that the gathering from the diagonal left anterior and right posterior winding lines has been carried to the point of discharge, which implies that it can be carried no further. We shall also speak of only a single S.

As regards the ball and the socket themselves, the attempt at forcing the *left anterior point of application* to repeat its course along the circumference of the left side of the socket will cause the pressure of the ball to slip over to the right side of the socket, by the rear—and when the *right posterior winding line* is subsequently forced, an attempt at a repetition of the effects of that line at its head-end will take place in the movement of the right side of the socket, producing an analgous result, viz. : the withdrawal of the socket on the left side by the rear, the counterpart of its previous action on the right side. Thus a cross-strain from the cross-tractions of the body will be preserved, but the lateral developments will be suppressed, their effects being carried toward the front (or in the snake, lower part) of the joint.

Similar effects of passing over a portion of the gathering to the alternate points would result *from the attempt to force the socket motions* brought on by the working of the diagonal right posterior point of application. And also even beginning with the points which correspond to the final movement of the primary sections of the winding lines, although these last would require a subsequent adjustment with the secondary sections, from having been carried out in advance of the regular course of movement.

It is the rotary movements (a) and (b), § 37, *i. e., those which accompany the formation of the secondary sections of the winding lines, that produce the arches. The movements across the joint from rear to front, (a') (b'), i. e., those which accompany the primary sections in their special movement, that depress the planes of the cross-lines and points of application.*

§ 73. Let us now look at the accompanying formations along the lines of traction. In doing this, we must bear in mind that the lines of traction are not permanent lines, but exist by certain

4

points of the skeleton being in such positions that the muscular actions play on them in certain ways. Thus, if the point where the anterior end of the cross-line for the right anterior winding line should be brought to a certain position in relation to the right anterior point of application, the preliminary form of the primary section of that line will be formed, and so on.

The *first* effect of *forcing the left anterior* point of application will be the passing over of a portion of the gathering from the secondary section of the left anterior winding line to what will be the course of the *right* anterior line. This transfer will begin, *first*, by the formation of a certain amount of convexity to the right, at the posterior end of the rear C, *i. e.*, allowing of the passage *without a spring* of a certain amount of the left side gathering.

Second. The *forcing of the movement* (a') will, at the same time, bring back the anterior cross-line end of the left anterior winding line to the centre and advanced to the front, and also cause the adjustment of the point, on the left side of the spine, which should serve as anterior cross-line end for the right anterior point of application, so that it shall come into a similar position.

Third. The *forcing of the movement* (*b'*). Here it will be remembered that the cross-line end being held in check (§ 32), the reaction to the right posterior point of application was at the head. The movement forward, across the joint of the socket will, therefore, first be equalized for both sides, and, then, the primary section drawing, the posterior end of the cross-line will be brought into place to the front, and the point which should serve as posterior cross-line end in the left posterior winding line be likewise similarly adjusted. *The requisite displacement for both now occurs to the front.*

Fourth. The *forcing of* (*b*), *i. e., of the outward rotation of the right side of the socket*, completes the adjustment of the cross-line ends, forms the left convex of the anterior C equal with that of the right side, and finally *equalizes* the reciprocal pressure of head-ball on both sides of the head-joint.

Remarks.—Referring to what was said in regard to the lines of traction depending for existence upon the relative situation of points in the skeleton, it is evident that when the *superimposition of twists* (§ 71) is completed, the diagonal lines may disappear, *and the tractions become collateral.* The diagonal relations, however, would be restored by very slight movements of the points.

§ 74. We have considered the right posterior point of application as actively working in the above movements of excessive action, but, since there is no spring, the induced action of this point by the continued working of its diagonal anterior point (the left) must be the actual course of the movement, and *we may regard the head ball action in* the left side of the socket as the only active one throughout. The secondary forcing point being active only in a sufficient degree to gather up in counter-action the line developed by the leading one.

In fact, *the difference between halting, by equalizing the four winding lines, as above, and locomotion, is that, in the first, the subsequent lines are formed by induction from the leading one, its action being the moving principle throughout; and as the effects are distributed through the frame, the first movement is continually repeated, while in the second the lines form independently and separately.*

Of course the snake may halt by merely ceasing the action with which it is moving, say just as it has come to the ground with the anterior point gathering, and when its shape would be that of the simple S S curves. We have selected the *third result*, not that it is the animal's usual way of halting, but, that it is the one important for our purpose.

§ 75. To retake diagonal gathering from the shape of "superimposition of twists," or, as it might be termed, "double diagonal gathering," it will not be necessary to repeat the diagonal workings of counter-action, for, since both gatherings are now present, the yielding of their lines of traction by one pair of forces to the other, or the overcoming of one pair of forces by the other, will restore the one-pair gathering on the slightest movement of the head.

§ 76. *The* "THIRD RESULT" *will form the* BASIS *of the system of* SETTING-UP.

§ 77. It should seem that the perfect locomotive actions of all animals are directed by the eyes.

Whether the eyes lead these actions, or only form a pivot for them, the steady point of every movement centres on the eye-pupil. It is necessary, therefore, that these poles of all the described curves should be freed from the necessity of accompanying any particular part in its movements. Such freedom is obtained partly by the appropriation of two or three vertebræ to form a neck, and partly by an *independent discharge of the* real or relative turning *movements of the eyeballs.* The discharge is *secured by a peculiar arrangement* which gives a steady appui to the ball, during the instant in which the small muscles that move it, discharge the twists received by them in accommodating themselves to the lines of traction of the body. The muscle on the inner side of each eye is made much longer than the others, and instead of simply fixing itself on the inner surface of the front of the eyeball socket, is there passed through a ligamentous loop, in which it slips freely, and then bending at a right angle, proceeds to the rear, to be fixed at the back part of the socket. The counter-turns received by this, *the " internal oblique muscle,"* are discharged by its slipping in the loop, while yet the eyeball is kept steadily in place by it. *This muscle may, when twisted, be regarded as an S, or as a rib, of which the eyeball is the foot end.**

§ 78. There is one more point in the anatomy of the snake which it is requisite to notice in connection with our subject, namely, the *Lungs.*

Although these organs be more imperfectly constituted in the snake than in the higher animals, they yet fulfil the function of expanding the chest

---

* See § 133 for detail. The relative action of the muscle remaining the same in the higher animals as in the snake.

so as to form an elastic cushion, compressible in any direction, but always ready to fill out again. On this cushion the front ribs rest themselves, and not only are the movements thus made more smoothly, but *actual aid is given to the ensuing gathering by the expansion of the air compressed in the lungs during the spring.*

We shall have occasion to go thoroughly into the discussion of that action of the anterior and posterior parts of the lungs, which in the higher animals belongs to the anterior and posterior portions of the central S curve of the body, and shall postpone further consideration of the subject to that time. We may, however, observe that the form which would give the *greatest capacity* to that part of the body in which the lungs are situated, is that of the *full gathered superimposition of twists* (§ 71). In this the ribs are raised toward the head by their ground ends as much as possible, while the point where the two turns meet is also raised outward by the depression of their spinal ends.

PART III.

APPLICATION TO THE HIGHER ANIMALS.

§ 79. Regarding, then, the snake's contortions as exemplifying the fundamental mechanics on which the locomotions of quadrupeds and man are based, we observe in these higher animals—

1st.—That the spine is not so pliable as in the snake; but that it is still a most elastic rod, made up of little blocks of bone, every two of which, instead of having, as in the snake, a ball and socket joint at their surfaces of co-adaptation, are, as it were, threaded together throughout the two surfaces by an infinite number of short, strong elastic filaments, which, rising from the whole surface of one block, run into the whole corresponding surface of the other, and form a solid elastic mass between them.

2d.—That from a number of the vertebræ the ribs have been removed. In fact, two such vacant spaces exist, one at each end of the spine, and between them is the space to which ribs remain attached.

3d.—That, on this space, the motion and the elasticity of the anterior ribs is very much reduced, in comparision with the posterior ones, and that, if the analogies are to be carried out, this reduction must be supplemented by the movement of other parts not present in the snake.

§ 80. The front space, bare of ribs—the neck—carries the head, and the head, as comparative anatomy has shown, is made up of consolidated vertebræ. Like the vertebræ of the back, *the head carries ribs, viz., the fore-limbs* (or arms), analogues of which have been traced, directly connected with it *and the lower jaw.*

It is easy, therefore, to conceive that the head may gather to itself, by means of its spinal attachments, of the connections of the lower jaw, and of those of the fore-limbs, all the threads of force collected by the trunk and limbs, and that, thus holding the moulds of all motions, it may direct their subsequent developments.

Thus the articulation of the head is the working medium between the brain and the body, when it gathers the ribs and limbs of appui on one bearing, with the eye diagonal to the rear appui, as the pivot, and when, completing this gathering, with the same eye as a point of direction, it double twists them on the other bearing, and holds the nascent curves of replacement ready to discharge their predecessors.

During these movements, not only do the *fore-limbs*, after moving with the body, finally settle into position with entire reference to the head, but the *lower jaw*, experiencing similar effects from the movements of the body, as conveyed to it through the lungs, finally clinches the whole by its reaction on the head.

From this condition of "*gathering*," in which the elements of motion have, so to speak, their orthographic projection on the base of the skull, the head, by the slightest positive or relative change in its bearings on the spine, may initiate, and by again changing these bearings, may complete any movement.

It is in this sense that we may consider *the head* to be *the governor of all perfect motion*. In deformed movement there is a failure of straight connection with the head at certain points, and this default of such connection must be made up for by extra and eccentric movements, which destroy the steadiness of the body.

§ 81. The hinder space, bare of ribs (the loins) carries, at its extreme end, a hollow but solid framework of bones called the *pelvis*. This consists of the two hip-bones, immovably joined together in front by the meeting of two bones called the *pubis-bones*, and as immovably connected behind by another bone called the *sacrum*.* These bones, though seemingly one mass with the pelvis in the full-grown animal, are originally distinct.

The *sacrum* is evidently an analogue of the spinal vertebræ, and we shall, further on, attempt to show that the *pubis-bones* represent a continuation of the breast-bone, the *hip-bones* standing for the ribs.

The sacrum is a direct continuation of the spine, and this, in quadrupeds, is again continued by the tail.

The pelvis is not connected with the spine by a free joint as is the head, but by the same sort of juncture that exists between the several vertebræ of which the spine is made up.

The whole pelvis, thus moving in one piece, answers, in locomotion, the purpose of an exaggerated vertebræ belonging to the trunk of the body as a whole, and furnishing an extended sweep for the thigh-bone sockets which it carries. *The trunk of the body may thus be considered as a* COMPOSITE SPINE, *of which the legs and, for a certain share in their motion, the fore-limbs are the ribs.*

---

* The pubis-bones are, at their rear ends, keyed to the hip-bones by other two bones (one on each side), called the ischium-bones, which in man form the seat bones.

§ 82. The ribs of the higher animals are, like like those of the snake, attached to the spine by ball and socket joints. The relative positions of the ball and socket, however, are reversed, the rib now carrying the ball; and *two* vertebræ, at the sides of their junctions, carrying the socket. Also, the further extremities of the ribs, which in the snake would have been their "ground-ends," are elastically joined on either side to a solid piece of bone called the *breast-bone, or "sternum."*

This bone, which represents all the ligamentous scutæ of the snake (§§ 44, 66) consolidated, may be considered as a *substitute ground* on which the ribs perform their movements, as those of the snake do on the real ground. At each spring the *breast-bone* is taken up and carried to the position required for the next effort.

Almost the whole length of this artificial ground is occupied, on both its sides, by the attachment of only part of the ribs, namely, of the first division from the head, called the "*true ribs.*" These are separately articulated to it—each rib by the end of its elastic prolongation.

Of the remaining division, called the "*false ribs,*" each rib has a longer and still more elastic prolongation than a true rib; but these prolongations, before reaching the breast-bone, become united on each side into a single *one*, and the two resulting continuations are attached to the hinder —in man the lower—end of the breast-bone.

The *true ribs* diminish in length and capability of independent movement as they approach the neck, so that the upper ones have very little beyond a hinge-like movement at the spine, as their front ends are raised when the breast-bone is drawn up by muscles from above, in connection with the outer turning of the false ribs below.

In man there are seven true and five false ribs; in the horse, eight true and ten false; in the lion and cat, nine true and four false; in the giraffe, eight true and six false; variations which, no doubt, favour certain peculiarities in the motion of each animal.

In this way the breast-bone answers, at its sides, to only a part of the length along which the rib sockets extend, being, in fact, only under those of the true ribs.

At its rear, however, the breast-bone is powerfully acted on by the false ribs, and these, principally, give it direction and push it forward, the false ribs of either side driving their motion through the breast-bone up to the head by a connection presently to be described. At the head it is received and adjusted with the action of the true ribs of the other side.

It is possible that we ought to assign one or two of the true ribs to the rear C of the spinal curve; but the proportionately greater length of that portion of the spine to which the false ribs are attached, and the fact that the first—or first two—upper ribs* seem rather to belong to the joint of the root of the neck (§ 90) and the spine than to the front C, might give sufficient preponderance of force as propellers to the false ribs alone (§ 62). *This being the case, we have adopted the more convenient nomenclature of assigning all the true ribs to the anterior C, and all the false ribs to the posterior one.*

§ 83. If we give an artificial ground to the ribs, one S in the spine will be required to work on it; if we attach two rear appui to the trunk thus formed, another S will be required for their support; and if the whole is to be centred on the head, still another S will be required for combining the two first, and this last S will require some separate appui on which to effect this combination.

These requirements, we think, are fulfilled in

---

* These ribs, we believe, are bent from front to rear in a different direction from the other true ribs.

the back or *dorsal* vertebræ, as the *central S;* the loins, or *lumbar vertabræ and sacrum*, as the S for moving the hinder limbs, or appuis *proper* of the trunk; the *neck* as the S of combination; and the *arms* as the appuis of combination.

We shall name these three S S S, beginning at the head, the *first S*, *second S*, and *third S;* or, the *neck S;* the central, and from its more simple action, the *ophidian** *S;* and the *loin S*.

The relative actions of these three S S S become quite changed from those of three successive S S S in the snake, as will be explained.

§ 84. The cavity formed by the ribs and the breast-bone is filled by the lungs, and from the extremities of the breast-bone go muscular cords of connection to the head and to the pelvis.

These cords of connection join the actions of the central or ophidian S, with those of the loin S, and the actions' of both these with those of the neck S.

The trunk, composed of the dorsal and lumbar vertebræ, the ribs and sternum and the pelvis, may be regarded as an S, of which the hinder limbs are the ribs. We shall call this *the composite S*. The

---

* Acting through the ribs, directly on its artificial ground, it has a true ophidian or snake action.

pelvis is a vertebra in it. Its posterior C is separated from its anterior C by the diaphragm, which will be described further on.

The head, neck, shoulder blades and this "composite spine" may again be regarded as a compound spine, of which the arms are the ribs. We shall call this the *bicomposite spine*. The head, taken with the shoulder blades, is a vertebra in it, distinguished, however, from the pelvis as a vertebra, in that while the "composite spine" governs the pelvis, the head governs the "bicomposite spine."

To some extent the arms act as ribs to the composite spine; also, and through them, the head draws in a direct way on the hinder limbs.*

§ 85. *Five vertebræ—two for each flexure, and one for the point of contrary flexure—are the smallest number of which an independent S curve could be composed. If, now, we allow one at each end for its articulations with other points, we have seven verte-*

---

* Although the head be the "governour" and the spine the proper "origin of all movement," movement may, in the higher animals, be *initiated* in other parts. For example, a man in dropping from a height may, by thrusting forward his hands or jerking back his elbows, and thus changing the centre of gravity by altering the shape of the body, through the medium of the breast-bone, change very materially the point on which he alights from what it would otherwise have been. . A horse in taking a fence, often does not know the ground on the farther side, and no doubt while yet in air, he can, to some extent, determine his point of descent.

bræ, *the unvarying number contained in the neck of man and* (with the exception of the sloth) *of quadrupeds.*

§ 86. The breast-bone is steadied at its upper part by muscles which—replaced by equivalents in quadrupeds—exist in their most simple manner in man. We shall, therefore, since we do not aim at discussing the minute differences of action which make an alteration in these muscles necessary, consider the type given in man as applicable for reference, whenever we speak of these, or indeed of any of the muscles.

In man these muscles,—from the upper end of the breast-bone, called the sterno-cleido-mastoids, or as we shall generally name them, the *sterno-mastoids*,—are very prominent, and may be clearly seen one on each side of the neck,* passing from the top of the breast-bone upward, backward and

---

* These muscles take their name in man from their various attachments, viz., each on its side, to the breast-bone, the collar-bone, and the mastoid protuberance of the skull, just behind the ear. In such animals as have no collar-bone the middle word of the name evidently falls away.

In the horse this muscle is replaced by two—" one, the '*sterno-maxillary*,' is fixed to the anterior end of the sternum, and passes up the front of the neck to be attached to the back part of the lower jawbone; the other, the *levator humeri*, is fixed to the front and upper part of the round bone of the shoulder, and by a detached slip to the upper end of the sternum, whence it passes up the front and side of the neck, to be attached to the mastoid projection of the skull, giving off, on the way,. slips which are attached to the four or five upper neck vertebræ."

outward to either side of the head, behind the ear. They are so attached to the bottom of the skull, near to and on each side of its articulation with the spine, that they either draw the head down toward the breast-bone, or the breast-bone upward toward the head, according as the resisting lines of the spine, as developed in the neck, are brought forward or backward between the two muscles, and so by their positions give the preponderating leverage one way or the other. That is to say, if the plane of junction of the two C C of the neck S be well thrust forward between the sterno-mastoids, the action of these muscles will raise the breast-bone; whereas, if this be retired, the same action will lower the head.

Suppose the chest well raised and filled with air, and the neck fully gathered by the "superimposition" of the diagonal counter-actions (§ 71), the plane of junction between its two C C will be advanced, and from this position every change of bearing in the head articulation will alter the drawing of the sterno-mastoids at the upper corners of the breast-bone, whether of both equally or of one preponderatingly. *Thus a perfect action of the neck is all important to perfect movement.*

§ 87. The *sterno-mastoids* must *partake of* the

double-twisting action shown for the internal oblique muscle of the eye (§ 77), although with them there is no necessity for a pulley attachment, the support against the neck-spine answering the same purpose.

It is easily seen that the two *sterno-mastoids* approach each other as they descend in front of the neck. Thus their *separate action must be diagonal*, ending, in fact, for each at the diagonal hip-joint, so soon as the intervening ophidian S has placed the points in the requisite position, by moving its substitute ground, the breast-bone.

§ 88. At *its posterior end* the *breast-bone* is *steadied against the pelvis by a combination of muscles*, which, in relation to the pelvis, carry out from the breast-bone a working in harmony with, but subordinate to that proceeding from the breast-bone, through the sterno-mastoids toward the head.

§ 89. In the snake a simple ball and socket joint at the head answers every purpose, for allowing the working of the spine over the ribs, these latter being the simple and only appuis; but in the higher animals another arrangement becomes necessary.

This consists in dividing the simple ball and socket articulation of the snake from behind forward, by a wedge-shaped fissure, and giving to each of the parts thus separated an ovoidal form. The articulation is thus divided into two separate working pieces, one used when the head-ball pressure is from the left, the other when it is from the right, and in the same way for the sockets.

Bearing in mind that the head joint always opens in front, we may represent this new form of the articulation by supposing the two halves, which would be obtained by the longitudinal division of a pear, about one inch long, to be soldered by their flat surfaces to the base of the skull, in such a manner that the points converge in front, while the globular ends diverge in rear, at an angle varying in different animals, according to the mode of action to be accommodated.

These ovoidal pieces are called the *head-condyles*, and have each a separate socket of corresponding shape on either side of the upper surface of the first vertebra of the spine.

§ 90. In the snake a couple of vertebræ, freed from the ground, are all that is required in order to adjust the eccentric movements of the anterior part of the spine with a steady position of the head; but, as we have seen in man and quadrupeds, there is, after what may be called the ophidian or snake movements on the ribs, another set of movements, namely, that on the hinder limbs, to be accommodated, and on this set again, in order to provide for the master movement of the fore-limbs—carried to its perfection in man—there is yet another set of movements to be accommodated, namely, that on these fore-limbs and on the lower jaw.

It is with reference to these requirements that we should explain the necessity for a complete S in the neck, added to the couple of extra vertebræ used in the snake; and, in dividing the vertebræ for this purpose, *we should place the one or two vertebræ answering to the snake's neck, directly under the seventh vertebra of the neck* (§ 82), *where they, aided by the general motion of the part, form a* VIRTUAL *ball and socket joint.*

The separate motions of the trunk and of the limbs may, we think, be easily noticed in the horse, and, indeed, Seeger, a Prussian author, in his "Horsemanship," insists much upon marking it. "First the body moves, then the limbs."

§ 91. In the chest of the higher animals, as has already been noticed, the ribs have great mobility and large development below, while at the extreme top they have little of either. It is apparent, on looking at this arrangement, that although the secondary sections of the anterior winding-lines (§§ 25, 68) might be easily established, from "points of application" acting on an anterior plane section, just below the first pair of ribs, and also the primary sections of the posterior lines, from "points" on a posterior plane at the summit of the lumbar vertebræ, yet the establishment of the primary sections of the anterior lines, and of the secondary sections of the posterior lines, would be seriously interfered with, unless the upper part of the chest be allowed considerable eccentric movement, such as shall successively bring its rib articulations under the course of the sectional lines referred to.

This eccentric movement is allowed and regulated by the action of the neck S, the lower part of which moves with the upper part of the chest, and must adjust itself by a virtual joint, in which the lowest neck vertebra and one or two of the most anterior ribs take part, and which we shall call the "*neck root joint.*"

The *head condyles* are here essential, in their character as two separate ball and socket joints, because of the break in the "composite" movement caused by the interposition of the breast-bone, which (§ 87) must place the intermediate points of traction in position before the head and pelvis act upon each other. This makes necessary an extra spiral-line movement, very much more extended than any for the connection of the S S, or C C, in the snake's contortions, and which cannot be accommodated on one surface. *It takes place, for the transition from the anterior winding-line of the "ophidian" to that of the "composite-spine," at the posterior end of either condyle;* and *for the similar movement between the posterior lines at the anterior end.*

Further, when the neck, S, allows this eccentric movement to the upper part of the chest, the line of pressure becomes oblique, and to resist this the condyle-sides must be rounded in the perpendicular plane. For example, the inner side of the left condyle to meet the movement of the upper part of the chest to the left, which accompanies the establishment of the primary section of the left anterior line, and its exterior side against the movement to the right, which accompanies the formation of the secondary section of the right posterior line, exerted against this side by the upward movement of the socket reciprocal to its withdrawal on the right §67 (*b*).

It is to be remembered that, although a part or all of these pressures may, at first, be *latent* in the neck, they must finally be satisfied at the condyles when the spring takes place.

§ 92. To illustrate what has been said:

Suppose the LEFT ANTERIOR WINDING-LINE to form in the OPHIDIAN S. Its *secondary section* will, to some extent, throw the left hind and right forefeet on their outer bearings, by the change in the centres of gravity of the body. It will also cause some movement of the head-joint, as if a simple ball and socket existed.

At the following combination of this "secondary section" into the "composite spine" (§ 84), the limbs of appui will be thrown decidedly on their outer bearings, and the relative motion at the head joint will be that of the posterior extremity of the left condyle revolving, to the rear and inward, in its socket.

Its primary section will occasion a further change of the centres of gravity in the same direction as at first, and, as it moves the anterior cross-line end toward the left side, will require a movement of the upper end of the chest in that direction, as well as one forward to correspond with the movement forward of the head-ball in the socket (§ 67 a′).

At the following combination of this "primary section" into the "composite spine," in order to keep the head steady, the above left movement must be met by resistance along the inner edge of the left condyle, and the movement forward by the moving of the condyle along this line.

Suppose next the RIGHT POSTERIOR LINE TO FORM IN THE OPHIDIAN S.

Its *secondary section* reflected action—the cross-line end being displaced, (§ 32)—will take its first re-action against the anterior plane section of the spine (in which the anterior point of application moved) bringing the posterior left corner of the sternum to the right, as the false ribs come on their inner bearings, and so far as it affects the upper part of the sternum, carrying it further to the left. The change in the centres of gravity will commence the change of the left hind and right fore-legs to their inner bearings.

At the following combination of this section into the "composite spine," the turning of the socket against the anterior end of the left condyle would occur, as the right socket is withdrawn; but as

(b') cannot take full effect until (b) is carried out (§ 37), this movement will be held in abeyance.

Its *secondary section* direct action will move the anterior part of the body to the left, and as it brings about whatever action may be equivalent to putting the right propeller true ribs on their inner bearings, the alteration of the centres of gravity will bring the left hind and right fore-feet still more on their inner bearings.

At the following combination of the "primary section" of the posterior winding-line into the "composite spine," which, in this order of succession, would be the final action in forming that spine, the diagonal feet of appuis will be brought strongly upon their inner bearings, and there will be required to resist the passage of the body to the right, as, reciprocally to the withdrawal of the right socket the left socket rises, a pressure of this latter against the outer edge of the left condyle, and corresponding with the suppressed advance of the socket (see the preceding paragraph), a passage forward, along this edge, to the anterior point, where, *finally*, a turn to the front and inward by the left socket will discharge the gathering, so soon as the alternate set of motions be inaugurated to release it.

§ 93. While the action of the ophidian and composite S S tighten the right corner of the sternum, and the right fore-leg against the left hind-leg, *a little additional motion of the head-joint tightens the left sterno-mastoid, and thus produces the "bicomposite S."* (§ 84). The bracing of the composite and ophidian S S is diagonal, that of the bicomposite is collateral.

§ 94. *In this gathering* movement *the front appui* (here the right fore leg) *goes with the body*, and, as we shall see, *the free fore-limb* (here the left) *principally and finally with the neck*.

§ 95. Each motion of the head condyles affects the sternum through the medium of the neck S, but its more immediate action, through the sterno-mastoids, so directly affects this bone, that if a head condyle (not an anterior point of application, at the front of the chest, and, thus the "bicomposite," not the ophidian S) begins the gathering, it will cause the sternum to move before the spine, a circumstance which, as we shall see further on, may explain those movements of the limbs—the pace, &c.—which do not seem to be strictly diagonal.

§ 96. In attempting to follow out the effects of the head condyle movements on the sterno-mastoid muscles, a very important clew seems to be afforded by the analogy of those muscles, which, in the horse, replace the sterno-mastoids in man. These are (see note to 86) *one muscle*, which goes from either upper corner of the sternum to the rear part of the lower jaw, so that its action will close the jaw ; or, this done, bring down the head toward the breast ; and *a second*, which goes from the mastoid protuberance, and the anterior C part of the neck S to the upper part of the upper arm-bone, the action of which is " to raise the shoulder and arm, and at the same time draw them forward, or, these being fixed, to turn the neck and head to one side."

We should conclude from these facts, as analogues, that certain effects, accompanying the motions given to the sterno-mastoids by the upper corners of the sternum, largely, even if indirectly, affect the motions of the lower jaw ; and also that the clavicle (collar bone) in man secures certain connections between the motions of the head and the arm-bone, and between the sternum and the lower jaw, which for the horse (it having no collar bone) must be supplied by extra muscles.

It would seem that when the *sterno-mastoid of the left side was drawn tight,* i. e., (§ 93), *when the* *bicomposite spine* (§ 84) formed, the lower jaw should be firmly closed and set on the left side.*

Also that, under the same circumstances, the left fore-leg should be raised, or if this were fixed, the head drawn somewhat to one side. This would corroborate what has been said (§ 94) in regard to the free fore-leg being under the influence of the neck.

Again, as the left sterno-mastoid must relax after the spring, so must the left side of the jaw, then the opposite (here right) side will begin to take appui in its socket, we expect to show how either side of the jaw gradually sets in its socket, and how, in bearing on its outer or inner side, its actions resemble, with reversed relations, the "bearings" of the ribs and limbs.

§ 97. Great as is the analogy as to shape which the fore-limbs have to the hinder-ones, this analogy is by no means complete. Both are modified ribs, but the action of the two is not only different, in that the fore-limbs act more largely as supporters, and the hind-limbs as propellers, but, while the hind limbs follow chiefly, the movements of the "composite S," the fore limbs, although connected

---

* We instance the left side, but, of course, with proper changes, the same holds good for the right.

with all three S S S, *finally* depend directly upon the head.

Thus, although supported by the anterior ribs, and adapting themselves to their motions, the *shoulder-blades* (which carry the sockets of the fore-limbs) have another connection, viz., that of the neck, and move on a basis exterior to the trunk, during the virtual crossing of the median line by those ribs. During this time, the neck is the S of whichever fore-limb is thus engaged. The quasi ball and socket motion, at the base of the neck, is the pivot on which either fore-limb changes from the trunk to the neck connection.

The *shoulder-blades* are connected in *rear* with the head by masses of muscle, and with the whole back ridge of the spine, with the ribs and with the pelvis, either directly or through the medium of the upper part of the arm-bone, by sheets of the same substance. To the neck part of the spine they are not directly attached, but are joined on either side by muscles to the whole length of a cord or ligament, which, loosely attached to the central rear line of the neck vertebræ, stretches over them all, from the head to the projecting bone of the lowest neck vertebra in man, and in the horse to the top of the withers.

In *front* the shoulder-blades are attached to the *breast-bone* and to the front of the ribs by large and powerful muscles, and the breast-bone being joined to the head by the sterno-mastoid muscles, there is, in this way, a *front* attachment of the shoulder-blades to the head. The massiveness of these connections agrees with the function of the fore-limbs as the final brace on which all the gatherings are collected.

§ 98. In the horse the breast muscles suffice for the connection in front; but in man the *collar-bones* are added. These are articulated, each at one end, with an upper corner of the breast-bone, and at the other end with one of the shoulder-blades.

The breast-bone of man is thus enabled to *push* against the arm-bones, as well as, like that of the horse, to draw upon them; and the head of man being, during the working of the condyles, thus braced against the "substitute ground" (§ 82), by the collar-bones, as the head of the horse is against the real ground, by its fore-legs, the supporting thrust is transmitted to the pelvis, and in this way the legs of a man can act both as complete "supporters" and complete "propellers."

§ 99. Thus, *first*, the *intermediate* appui of the action of the neck is the sternum, on which the action of the neck follows, nearly to completion, the torsions of the loins and of *the ophidian S*, making the *centre* of this latter *between (the lower points of the shoulder-blades,*) the centre of force,*† the *neck*, being itself *the centre of action*. Then another torsion, in which the eyes participate, and into which are brought the final turn of the limbs, together with the finishing of all the torsions, makes *the head the governour of direction*.

§ 100. It may be remarked that in making the foregoing distinctions of three S S S and of simple composite and bicomposite spines, no one of these, in a perfect body, acts without, at least in some degree, affecting the others. Thus, the slightest change in the ophidian S should alter the line of gravity over the feet, and require a motion of the whole frame to adjust it.

§ 101. In man the *five toes*, the *five fingers*, and the *bones* which *in the palm of the hand and in the sole of the foot* support them, would seem to *represent* the *ends of five ribs* belonging to five vertebræ, required to make up an S (§ 85), and, as such, we

---

* Where, as will be seen further on, is also the centre of the lungs' action.

† This fundamental centre not being reached, unless the working of the body is perfect, even a moderate distortion by the right-hand deformity occasions a great loss of power and accuracy in all motion.

may suppose them to be attached, the toes to the composite—the fingers to the bicomposite—spine (§ 84). The bones of the arms and legs would then be the consolidated masses of these sets of five ribs.

To avoid confusion, where speaking of man and animals under the same head, we shall call both fingers and toes "*Digits*," and number them beginning with the thumb or great toe, *first, second, third, fourth* and *fifth* digits.

If the third digit ("middle" finger or toe) represent the centre of this S, the fourth and fifth digits will receive their impressions from the rear C, the first and second from the front C, and the central one from both. Indeed we may go further and say that the fifth digit represents the posterior part of the rear C, the fourth the anterior part, the second and first the same for the front C. Then (§§ 25, 92, 96) the *fifth digit* receives its impressions from the formation of the secondary section of the anterior winding line, the *second* from the establishment of the primary section of this line, the *fourth* digit from the establishment of the primary section of the posterior winding line, and the *first* from the formation of the secondary section of that line. They will, also, in these connections be re-

5*

lated to the actions of the head condyles (§ 92) and to those of the lower jaw, which latter will be more fully considered hereafter.

§ 102. It will be observed that, regarding the fore and hind limbs as ribs, the counter torsions in their length can no longer, as in the real ribs, be received by the elasticity of their substance; accordingly, separate joints are substituted for elasticity. Again, were the bearings mere bevels (§ 51), the passage from one to the other would be very rough; and, unless the bevels were very broad, quite insufficient for the extent of motion. This is remedied by the formation of the foot of the higher animals. In these, however, it is unnecessary that the first digit appear as an actual appui, if they rest on four feet; it is only in man, and, to some extent in a few other animals, that the development and spreading inward of the great toe renders possible the steady change in the S S, as their forces cross the median* line without any support from anterior appuis.

To still further assimilate the perfection of support to that which the snake gains from a completely underlying set of appuis, it is necessary that the

---

* Any shoe which interferes with this inward spread of the great toe is a hindrance to marching.

digits supporting the motions of the rear C should still remain on the ground, while those supporting the front C are being brought in action. This seems to be accomplished for the hind limbs by the projecting heel, which, by means of the strong muscle passing from it to the thigh bone, allows of the foot being rotated on to its outer side, while the passing line of gravity, at the same time, brings its inner side down,— or vice versa. The elbow in the horse may answer the same purpose in respect to its fore limbs ; while in man the power of turning at the wrist should seem to make any other appliance unnecessary.

The heel muscle of the hind limbs subserves, of course, other purposes ; being one of many which give to the limb, and that with vastly increased force, all the elasticity of the most elastic ribs. *

---

* It may be doubted that the heel of a man should, in perfect locomotion, touch the ground at all; but the idea of this being an essential part of the step has caused an ingenious writer, and apparently capital horseman, Captain Raabe, of the French cavalry, in his work " Examen du Cours d'Equitation de M. d'Aure," 1854, to suppose that each foot of a man, when walking, goes through the motions of the gallop of a horse, leading with the left leg for the right foot, and with the right leg for the left foot; that is to say, the outer edge of either heel stands for the horse's outer hind foot (1st. beat) ; the inner edge of the heel and the outer toes, for the inner hind and outer fore-foot (2d. beat) ; and the great toe for the inner or leading fore-foot (3d. beat). He says the change of the leading foot, which can thus be performed in place at each step by man (as in "mark time"), was never performed by any

§ 103. In the *bear*, &c., the pieces of the foot are still five, and all of them are still applied flat on the ground. In the *dog* they are reduced to four; the digits themselves are still put flat on the ground, but the bones (analogous to the palm of the hand in man) which carry the digits, are raised upright. In the *horse, ox,* &c., not only these, but all three sections of the digits (these sections may be counted in the joints of a man's fingers) are also set upright, and the animal moves, as it were, on its nails which have now become "hoofs." At the same time, in these last animals, the bones which carry the digits have been very much lengthened, and form the "cannon." In the horse, the third and fourth digit (as we should suppose) have been consolidated into one, and form, by their joints, the "large and small pastern bones" and the "coffin bone." In the cannon of the horse's leg, the third and fourth of the bones which carry the digits, are consolidated into the

---

horse, excepting "Partisan," an animal, by subduing which Baucher founded his reputation.

Thus, Captain Raabe seeks to establish the high character of the human walk, and to refute the slander which, he adds, has represented it as a sort of broken down amble. Our description would count the heel as a support only when supplying the failure of the outer toes to perform their functions, and, therefore, cannot fit in with his supposition.

"cannon-bone" and the second and fifth remain as the "splent-bones." These last do not reach down to the length of the cannon bone, and they no longer have any digits to carry. In the horse and in the dog, the first digit is, at the most, *represented* by some dislocated piece.

* The hock-joint is formed of the small bones, which, in man, compose the ankle; and the "point of the hock" is the projecting heel of man. In like manner, the small wrist bones are all found represented in the "knee" of a horse or ox, while in these animals the elbow rests close to the body.

§ 104. In the bottom of the horse's foot, two lobes, divided by the cleft of the frog, may be easily seen. Now, if we suppose two fingers of a man's hand to be placed together, the ball of one finger against that of the other—next, that while the nails grow together *between the fingers, while on the outer sides* the nail of each outer finger turns and prolongs itself into the skin—we have only to draw away the skin from the inner face of these nails, to get a representation of the bottom of the horse's foot; *the* NAILS *and their inflexed continuations standing for the "crust" and "bars;"* the SKIN *outside the turning of the nails, and what is drawn back from their inner surface, for the "frog;" and the portion of skin stretched by this drawing back, for the "sole."*

If the foot of the horse contains, as we have supposed, only two digits, we must look for the point of effect of the secondary section of the "anterior winding-line" (§ 101) to the outer splent bone at the hock, and regard it as only *mediately* affecting the foot, through the connection of this splent bone with the cannon bone. In the same way, we must find the point of effect of the secondary section of the posterior line on the inner splent bone, and its action on the foot only *mediate*, as before.

The action of the secondary section of the posterior lines and the *final adjustment*, which, in man, would be carried through the *thumb, in imaginary appui*, must, for the horse, depend on movement in the fore legs, and in the quasi ball and socket at the base of the neck (§ 90), an unusual requirement for this movement may possibly account for the "dishing" of a raised fore foot in some horses. The dishing is evidently a prolongation of its inner bearing, and this prolongation may, perhaps, be owing to uncommonly extended action of these parts before the "lift."

The *ox*, &c., have two parts to the hoof, and, if

these parts represent each the consolidation of two digits, i. e., the second and third in one, the fourth and fifth in the other, the fact of there being but one bone in the cannon of these animals may show that all the bones having, in this case, digits to carry are consolidated into one for action upon them, and corroborate the supposition that the splent bones in the horse remain separate as the unemployed connections of two discarded digits in its foot.

§ 105. It will be well, perhaps, before giving a description of the diaphragm to make some allusion to *the lungs*, although a more detailed examination of their functions will be made further on.

It has been remarked (§ 76) that when the snake's ribs are fully under the influence of the double turns given them in the "superimposition of twists," then the chest has its greatest possible capacity, and that as a consequence of this the *lungs*, which fill the chest, are then expanded to their greatest limit.

If this air be forcibly detained within them during action, the *lungs* will form *an elastic cushion*, which expands after every compression. They thus, like the fly-wheel in machinery, make good any deficiency in force from other sources at every part of the movement. Compressed by one gathering they aid in initiating the next; supporting the chest they form, AT THEIR WORKING CENTRE, (COMMON WITH THAT OF THE OPHIDIAN S) THE TRUE CENTRE OF FORCE (§ 99); and they fulfill another and more important function in the *completion of each movement of locomotion*, which will be spoken of in connection with the lower jaw.

§ 106. Within the ribbed portion of the trunk in the higher animals, are the lungs, and below this portion the stomach, bowels, &c. The two are separated by a sheet of tendon stretched horizontally across the bottom of the chest and attached by muscular fibres to the upper edges of the lowest ribs, and to the cartilage-prolongations which go from them to the lower end of the breast bone.

This sheet of tendon with its muscular border is called the DIAPHRAGM. In our description of it and of the muscular fibres which stretch it we shall speak in general terms, aiming simply at conveying such an idea of its action as may be of service in explaining the process of "Setting-up," when we come to that final object of our work.

The diaphragm is described in anatomical works as consisting of three lobes, whose shape and situation may be represented by a *trefoil*, having its stem fixed to the spine.

The muscular fibres form a border from the edges of the two tendinous *side lobes* to the lower rim of the chest. They will, of course, stretch the diaphragm, when the ribs turn so as to favor this action, viz., concave surfaces of ribs to the front, i. e., the outer-bearing. The muscular fibres from the front part of the *central lobe* going toward the breast bone will complete this stretching, when the heads of the ribs, by turning in a contrary direction, i. e., on their inner-bearing, and sinking back into their sockets shall give a double twist to the ribs.

So far, the *movements of the diaphragm are governed by the secondary section of the anterior winding-line, as to the outer bearing, and by the primary section of the posterior line as to the inner bearing.*

The *primary section* of the anterior line simply increases the effect of the secondary; but the *secondary section of the posterior line* which (§ 25) gives inner bearing to the upper C C of the S S, *has its action* in stretching the diaphragm, *aided by* additional muscular fibres. These form *two* long, thick *muscles* which, between them, gather together the tendinous fibres from the whole sheet of the diaphragm at the rear, as if the "trefoil" stem were split in two. They then pass, the *one on the right, the other on the left* of the upper vertebræ *of the loins*, and are fixed, as they descend, to those vertebræ which we should consider as *forming the upper C of the third S of the spine* (§ 83). From these vertebræ, as points d'appui, their action gives a final stretching to the diaphragm, in correspondence with the final inward bearing which is given to the whole body by the action of the secondary section of the posterior line, when, in the bicomposite spine (§ 93), it accompanies the motion of the condyle socket along the outer edge of the head condyle. These muscles are called the "*Pillars of the Diaphragm.*"

§ 107. The diaphragm supports the lower surface of the lungs, and thus, from the double-twisting action of the spine, which draws the diaphragm flat, as well as extends the ribs, we have the chest expanded in two directions, length and diameter.

It is obvious that the above arrangement of the diaphragm is fitted to act in two parts, a right and a left. These conjoin their actions when the "curves of superimposition" are in force.*

---

* It may be added that, *under* the diaphragm lies, on the left side, the stomach, on the right, the liver. Any habitual enlargement of

§ 108. Beginning their upper attachments on each side, behind and parallel with the lower attachments of the pillars of the diaphragm, that is to say, on the upper C of the third S, are two long and thick muscles called the "*Psoæ*." They deverge from each other obliquely, outward and downward, until, passing from the inside of the pelvis, they reach the thigh bones, to each of which a psoas muscle is so fixed that by its contraction the thigh bone is rolled outward. The *psoæ* muscles, as we should suppose, *turn with the effect of the primary sections of the anterior winding lines on the upper C C*, and, with the "pillars," complete the analogy of the diaphragm to the "digastrics," two small but focal muscles to be presently mentioned.

§ 109. The various muscles above referred to mark out, we think, the leading lines by, and on which the curvatures of compression and extension are formed. They are assisted by a multitude of other muscles, some larger, some smaller, which all work in harmony with them, if the frame be undistorted.

§ 110. As centres of formation for the new curves,

either may, by interfering with the working of the diaphragm, induce distortion. Such trouble with the stomach would, by enlarging and *fixing* the course of the secondary section of the left anterior winding-line, favour the right-handed deformity.

as well as actual workers in giving the final turn in discharging the old ones, two pair of small muscles of very peculiar construction are especially concerned.

One pair of these, the *Internal Oblique Muscles of the Eyes*, has been already described (§ 77), when speaking of the snake.

The other pair, the *Digastrics*, through which the lungs are suspended by one end of each muscle from the lower jaw, and, by the other end, from the base of the skull, are, we believe, found only in the higher animals. We shall describe them in connection with the wind-pipe (§ 114), the action of which, in the locomotion of the body, these muscles may be said to express.

§ 111. The *Lungs*, which may be described as an "*air sponge*," but one absorbing from the interior instead of the exterior surface, are enclosed in cases which allow the air to enter and to leave them by only one and the same opening.

They are made up in man and the higher animals, so far as our purposes are concerned, of four* such inclosed portions, two on each side.

The chest, as can easily be observed, has a coni-

---

* There are, in fact, five such parts, three on the right side and two on the left. This disposition may have reference to the heart, which is on the left side.

cal shape without, and, within it, the lungs taken as a whole form a sort of cone, the apex of which is above, while the base rests on the diaphragm below.

The shape of the four incased parts, which are put together to form the cone, may perhaps be best given by supposing first, a perpendicular plane passed lengthwise through the spine and the breast bone, this will divide the lungs into right and left halves. Then another plane, oblique from above downwards its upper surface facing to the front, passed between the true and the false ribs, will divide the whole into four pieces, which are called "*lobes*," and the general shape of each of the four will be given by the direction of the planes.

§ 112. To each of the two upper lobes of the lungs appendages are added, at their upper ends, which may be called "*tips*;"* when unfilled these lie somewhat bent and twisted, at the summit of each lung. As we suppose every part of the lung to fill, not always in the same order, but according as the movements of the chest create a vacuum, these *tips* would, on our theory, *fill from above downward*, i. e., in the opposite direction to the filling

---

* That these "Tips" may have some special action, is alluded to in an article in " Townsend's Cyclopædia of Anatomy."

of the bodies of the upper lobes to which they are attached. They take, we conceive, a double twist, each for the action of its own line, and constitute the basis for the connection of the neck-action with that of the central spine, at the quasi ball and socket joint of the "neck root."

It will be observed that, as the space of the false ribs, ending in a single continuation in front, § 82, is more extended in rear than in front, so *the two lower lobes of the lungs* marked out by them will be very much *thicker behind than before*, while the *two upper lobes* will be somewhat *deeper in front than in rear*. The true and false ribs and the lung lobes thus seem to match in shape as they do in motion.

§ 113. The lung lobes are filled with air through only one set of tubes; but *where* (that is, in what part) they become filled depends upon where a vacuum is created in the chest, by the motions of the ribs and of the diaphragm. Experiment will —so we think—easily show, that *in the fillings from the ophidian movement the upper lung lobes fill from the lower part upward; the lower lobes from the upper part downward.*

There are *two moments of filling* for each set of lobes, *viz.: for the lower lobes,* when the secondary sections of the anterior winding-lines throw the ribs on their outer bearings, and again, when the primary sections of the posterior lines bring them on their inner bearings; *for the upper lobes, similar* moments, *with the necessary substitutions* as to sections.

Both these fillings will, for the ophidian action, be as just stated, but *at the time when each movement coalesces with the composite spine, the lung tips* (§ 112) will accompany the movement of the quasi ball and socket at the root of the neck, so that *whenever a connection of the lower C C occurs* the *lung tips fill at their upper part from above;* and *whenever of the upper C C, at their lower part, also from above.*

The formation of the *bicomposite spine* tends to *straighten* the *lung fillings*, destroying the counter turns, and making the whole lung one; entirely so, when the "superimposition of curves" (§ 71) has place, and approximately so in single diagonal formations.

There is a certain action at the lower edges of the lungs (about the diaphragm) analogous to that at the lung tips, but it seems unnecessary to take this into account.

§ 114. The lobes of the lungs consist of an infinite number of air-cells, which communicate with the air by a multitude of tubes; these unite and reunite, until they are reduced to one on each side, coming off about the centre of the surfaces of contact of the lobes of that side, and these two again unite into one principal tube, the "WINDPIPE." This sole conduit for air into the lungs can easily be felt as it passes upward above the breastbone.

§ 115. In front of the neck vertebræ the windpipe is surmounted at top by a sort of box, of considerable size, called the "LARYNX," prominent in front as "Adam's apple." *To this box there is but a single small opening*, through the membrane which forms its upper part, *and this opening can, at any time, be closed by a lid called the* "EPIGLOTTIS."

It is evident that the amount of air surface which can come into action against the small under surface of this lid is exceedingly limited, compared with the surface presented by the air in the lungs, and that thus the slender muscles, which close the lid, are quite competent to hold it tight against any escape of that fluid, even when the most violent compression of the lungs, from the action of the ribs, reduces the cavity of the chest at the spring. The epiglottis, still refusing to allow the escape of air, it is condensed, and the equivalent expansion assists in developing the alternate shape, so soon as the compression is removed.

§ 116. It is our next object to trace the succession in filling or exhausting the different parts of the lungs, as the one or the other accompanies the formation of the winding-lines, and their reaction on each other in the different C C.

Supposing, as usual, the left anterior point of application to commence the workings, and this by the ophidian S. The formation of the right convex of the upper C, and of the left convex of the lower C, by the left anterior winding-line, will enlarge the right upper lung lobe upward, and the left lower lobe downward (§ 113).

The reflex action from the *secondary section* of the right posterior line (b' § 58) very particularly enlarges the left lower lung, as it draws downward the rib articulations.

Meantime, the air is drawn from the left upper and right lower lobes and into the left lower and right upper.

Next we have the direct action of the secondary section of the right posterior line, coming from the trunk toward the articulations of the ribs (§ 58), and meeting the already formed primary section of the left anterior line. This completes the filling of the right upper lobe for the ophidian spine.

But the actions of the upper lobes are very different from those of the lower ones, by reason of the eccentric movement of their upper ends, for although there must be some eccentric movement of the lower lobes, from their being connected with the upper end of the third S, yet this is met by the considerable extent of the lower lung surface and the elasticity of the lower ribs. On the other hand, the upper points of the upper lobes must revolve with the ball and socket motions at the base of the neck.

Each one of the above ophidian movements is followed by the action which unites it with the composite spine, and which increases still more the filling of the lung, by rendering tenser the enlargement of the chest cavity at the same time that it fills the lung tips (§ 112).

Finally, the formation of the bicomposite spine in reducing the counter directions of the filling of the lobes and of the lung tips, expands the chest still further.

The *compressions of the lung are caused by the*

development of the new curve, and, of course, *are greatest at the moment of discharge,* when the shape of the spine changes.

§ 117. In connection with these movements, we think one may trace the relative workings of the *psoæ muscles* and the *diaphragm*—of the small *oblique muscles* at the back *of the head* and the *sterno-mastoids*—of the *muscles which from the upper part of the shoulder blades, converge at the back of the head,* of the *muscles which, from the back, pass to the arms,* and of the *muscular combinations which,* acting between the lower end of the breast bone and the collateral sides of the pelvis, straighten the tractions of this bone from diagonal to collateral.

In tracing the action of the bony skeleton, as affected by these muscles, we shall leave out of mention the Latissimus dorsi, the oblique muscles of the head, and those connecting the shoulder blades and the head, because we think that their workings may be more clearly defined by tracing the shoulder blades as resting on and following the shape of the chest, and that the movements of the head are sufficiently clear of themselves.

§ 118. Taking a general view of the action of

the three spines*—ophidian, composite and bicomposite (§ 84)—it will be observed, at the head ball and socket joint, that the result of a ball and a socket action, fully carried out, is to, as it were, *throw off* the point of pressure and the pair of counter-acting lines on which it depends to the front, disengaging them, one from the other, each to its own side. And if, as in the curve of "superimposition" (§ 71), there be two pairs of counter-acting lines, both are thus thrown forward and disengaged latterally. In either case, if not discharged, they will straighten, and, in the case of one pair, this position protracted will bring the other pair into action, and, suppressing the lateral developments, will produce double superimposition (§ 71), *and make the tractions collateral* instead of, as before, diagonal (§ 71).

Now, before the whole course of the lines for all the spines are thus made collateral, this ball and socket action occurs at four points:

First.—Between the upper and lower lung lobes, for the ophidian S.

Second.—At the diaphragm, for the union between the central and third S.

---

* When considering the bony spine only in relation to its S S S, we have called them First, Central and Third.

Third.—At the digastrics, for the union between the central and neck S, covering, to some extent, also the third S.

Fourth.—At the head joint, for the formation of the bi-composite spine, covering the whole.

·The second and third are those by which the two ends of the composite S are formed.

The fourth and first unite in their action after the action of the third and second, so that the central focus is always between the lung lobes.

The action of the digastrics, in representing the central uniting with the neck S through the neck-root joint, is similar to that of the diaphragm in uniting the central and the third S, the flat anterior part of the diaphragm (§ 106) being the analogue of the anterior portion of the digastrics, the diaphragm pillars of the posterior portions. The psoæ complete the action, and join it with the internal obliques of the eyes.

The various S S of the snake are discharged in succession; those of the higher animals are collected together under one S—the "bicomposite"—by means of the neck and then discharged together.

§ 119. The above is the gathering movement, and determines the action of the legs of appui. The free legs are moved in the discharge. In simple diagonal locomotion, such as the trot of the horse, the gathering of one pair of lines, and the discharge of the other pair, are to some degree synchronous, and must be described together.

We shall next enter into a more detailed examination of the action of various parts.

§ 120. The long tube, called the *windpipe*, is, during locomotion, the subject of various twisting movements, coming from the motions imparted to the lobes of the lungs by the varying shapes of the chest. These twistings are not communicated to the larynx, or voice box, because the windpipe, just below the plane at which the larynx crowns it, is held up by two loops, one on either side, and these slip on cords stretched from the back part of the under surface of the skull behind the head condyle joints, to the inside-front of the lower jaw, on a line between the two central incisor teeth. These cords, which are each composed of two muscular masses, joined together in the centre where the loops run by a smooth, round tendon, have been already mentioned as the *Digastrics*. Being joined to the skull *behind* the head condyles, to the lungs by the windpipe, and to the lower jaw, they obviously connect the motions of all three.

§ 121. The larynx is a box with unyielding sides, but the *windpipe* is a flexible tube, formed, at its front, by a series of three-quarter hoops. The space between the ends of these hoops, that is the back of the windpipe, is filled out by a yielding membrane. The windpipe thus somewhat resembles the body of a snake, if we suppose the ribs to join below, forming one piece with the scutae (§ 44) and the vertebræ to be removed above.

§ 122. It is plain, we think, not only that considerable motions of torsion and counter-torsion must take place in the windpipe, but also, that its connections with the lower jaw, to be given in the next section, show that those twistings must exert a powerful reflex influence on the motions of the whole body. We might also appeal to horsemen, who well know how different the effects of the bit are from those of the cavesson, a difference which can be explained only by the fact that movements of the lower jaw affect the lungs, and changes in the lungs affect the trunk. It is when the spur, acting on the false ribs near the diaphragm, alters the filling of the lungs, in counteraction to any attempt of the lower jaw to stiffen itself into one piece with the head, that that perfect mastery over the horse's motions is obtained, which is one grand object in cavalry riding.

§ 123. The larynx, windpipe and lungs lie, as has been mentioned, entirely loose within the chest and neck, and are *suspended* by the digastrics. Before speaking further of the lower jaw we will describe more particularly the points between which the digastrics are stretched. Just behind the ears may be felt, on each side, the bony protuberances to which the upper ends of the sterno-mastoids are fixed. Close *inside* of each protuberance is the attachment of a digastric. From these points the digastrics go forward, converging toward each other, and having run each through one of the loops on either side of the windpipe, they meet together and fix themselves inside of the front part of the lower jaw, on either side of a line passing downward from between the two centre front teeth. The mastoid protuberances, the centre of the lower jaw, and the top of the windpipe, are so easily to be traced, that after mentioning the fact that these cords *sag down* in the middle, where the windpipe is suspended, it seems unnecessary to dwell longer on their description.

§ 124. Just in front of each ear one of the hinge-like,* articulations of the *lower jaw* is easily recognized on opening or shutting it. The weight of the lungs and the contractions of the digastric muscles would continually open it by drawing on its front part, were it not for several powerful muscles on both sides which easily keep it closed, *so long as the windpipe loops do not come too far forward.*

On the plan of describing only such parts as seem necessary to the explanation of our subject, we may here select the two " *Temporal muscles,*" one on each side of the head, as representing the forces which close the lower jaw, and which thus, as will be presently explained, complete the series of motions that give a double twist to all the springs of the body.

For the lower attachments of the temporal muscles a projection of bone rises from the upper edge of the lower jaw on either side, in front of the hinge-like articulation with the skull, and, from these projections, the muscles spreading out like a fan, fix themselves for their upper attachment over the surface of the *temples* (whence their name).

§ 125. When the lower jaw is closed, each *di-*

---

* In the lion, &c., they are complete hinges; in man and the horse there is also a lateral movement.

gastric has effectually two attachments to the head, one in front and one in the rear of the head condyles, so that it is easy to see how the movements of the lungs and of the head are connected by these muscles in harmony with the workings of the neck. It is (§ 118) *through their medium that the shoulder and hind-limbs are brought into connection with the bicomposite spine.*

We have (§ 121) referred to some resemblance between the rings of the windpipe and the ribs of the snake combined with its scutæ (§ 44), supposing the vertebræ to be translated to some other position. *The windpipe* seems to act in sufficient correspondence with the vertebræ of the neck to admit of the supposition that it is *the " artificial ground" on which the neck works in the composite spine, as* the ribs do on the sternum, and, by the connection of the abdominal muscles, likewise on the pubis. The sternum is joined to the head by the sterno-mastoids, *just where* the windpipe is joined to the head by the rear portions of the digastrics. This fact may somewhat corroborate their similarity of functions.

The digastrics, when the lower jaw is closed, give the windpipe an attachment to the head, in front of the head-condyles, but the trunk of the body has such attachment only through the chest being raised by the lungs, and consequently, also only by the digastrics. We should, therefore, finally, consider the *lower jaw* as the *artificial ground of the bi-composite spine.*

§ 126. To sum up what has been said in the last and several preceding sections we shall have—

For the DIAGONAL ACTIONS the central, the lumbar, and the neck S S S.

For the BALL AND SOCKET MOVEMENTS connecting the diagonal actions and their lines of traction forward each to its respective "substitute ground" and from these to the real ground:

The *centre of the central S* between the lung lobes (connecting the two C C) making the ophidian spine.

The *diaphragm*, between the central and lumbar S S, making the lower half of the composite spine.

The *neck-root joint* between the central and the neck S S, making the upper half of the composite spine, the movement of which is represented by the *digastric muscles*, between the composite spine and the head.

The *head joint* forms the bi-composite spine, and through the shoulder-blades and lower jaw, so connects the parts, that all return *again to the movements between the upper and lower lung lobes.*

Finally, the *internal oblique muscles of the eye* throw off all the gatherings for a renewal of the actions, by the alternate movements.

The *ball and socket* movement, *at* the *root of the neck,* being at the plane of the points of application for the ophidian movement, and between the composite and bi-composite spines when their last movement of union takes place, *works* like the head condyles, *partially* with all the ball and socket actions, and ceases to do so only when these centre on the plane between the upper and lower lung lobes.

EACH OF THE SPINAL MOVEMENTS THUS PASSES TO A " SUBSTITUTE GROUND " to which they transmit the tractions so changed by their passage through their respective ball and socket joints, that, in the case of the synchronous action of all the counteracting lines, they become *collateral* just before the discharge, and in the alternate action by pairs, a like collateral union has also place at some instant during the step. These " substitute grounds " are—

Of the ophidian spine, the *Sternum.*

Of the lower half of the composite spine, the *Pubis*.

Of the upper half of the composite spine, the *upper part of the Sternum*.

Of the composite spine, which includes *partially* the neck (as it does *partially* the shoulders), the *Pubis*, *Sternum* and *windpipe* (§ 125.)

Of the bi-composite spine, the *lower jaw*.

For the whole body, the *eye balls*.

§ 126. (a) *The internal oblique muscles of the eye, occupying the same relative position as in the snake, are, first of all, and fundamentally connected with the ophidian spine. They accept the additional connections as they form, and, finally, by the rather extensive movement of the digastrics and the neck, the head joint being the centre, the shoulders and lower jaw are adjusted with them, and the bi-composite spine is formed.* This is the course for what we consider the fundamental action, viz., the TROT; certain variation taking place for the other gaits.

PART IV.

Locomotion of Man and of the Horse.

§ 127. Locomotion in man and in quadrupeds, although *apparently* so dissimilar, is yet, in its chief points, really so alike, that after discussing the action of the horse it will need only a few additions in order to adapt the description to the human frame. And *the locomotion of the horse* being also the nearer step in the transition from that of the snake, we shall commence with it.

§ 128. The trot, for reasons which will appear as we proceed, is more fully based on the ophidian action than any of the other paces, and will therefore more perfectly fit on to the discussion of the various tractions given in the preceding sections. On this account, we select it as *the fundamental gait* with which we shall compare all the others.

The trot not only exhibits the ophidian action in the change of bearing from the outer to the inner sides of the feet employed as appuis, (equivalent to the change of bearing on the ground end bevels of the snake's ribs, but the movement commenced in the chest at the ophidian S takes, as in the snake, a *nearly* synchronous spring from the diagonal points d'appui.

In this gait it is to be noticed that two diagonal feet being on the ground and the other two being lifted, the fixed fore-foot of one side leaves the ground just in time for the *collateral* raised hind-foot to be put down in its place—that on the opposite side the two feet (a raised fore and a fixed hind-foot) separate widely, the one forward, the other backward—and that, as we think may be recognized by the eye of an observer, the propelling hind foot leaves the ground *an instant before* its diagonal (the fixed fore-foot) does so. This result seems analogous with the view we have taken of the order of discharge of the two C C forming each S in the snake's locomotion.

§ 129. Let it be remembered that the air in the lungs expands to support every expansion, and condenses to allow every contraction of the chest; also that these air actions are of two kinds, *one central* in the lobes proper, coming from the ophidian movement proper, *the other terminal* in the lung tips, coming from a working added to that of the central S, in order to accommodate the action of the two additional S S, i. e., of the loins and of the neck, which work with the hind legs and the fore legs (or arms) respectively and to serve as a *working pivot* for the union of the ophidian and composite spines, as also by a repetition and further extension of its movements for collecting all in the bi-composite spine.

The *filling of the lung tips* (§ 112) takes place as if each lung-tip were an S. The upper part, or C, fills from above with an outward turning motion, harmonizing with the outward turning of its diagonal lower lung lobe as this filled, and, by a further outward turning, not accompanied by filling, harmonizing with that part of the filling of its attached upper lobe which requires an outward turning. The lower part, or C, fills from above with an inward turning motion, harmonizing with that part of the filling of both the attached upper and diagonal lower lung lobes which requires an inward turning, but especially with that of the attached upper lobe, with which this inward turning finally makes one piece of the lobe and its tip.

It will be borne in mind that the filling of the lung lobes is accomplished in two times—(1.) By the secondary section of the anterior winding-line for the lower lobe—of the diagonal posterior winding-line for the diagonal upper lobe, causing an outward turning. (2.) By the counter-action of the two lines causing an inward turning throughout their points of meeting, which, finally, raises the outermost point of the lower edge of the ribs and gives the greatest capacity to the chest.

The filling of the upper C of a lung tip is connected with the formation of the upper part of the composite spine in the anterior winding-line. It thus accompanies the turn outward from the lower part of the collateral sterno-mastoid and of the diagonal diaphragm pillar and psoas. The filling of lower C is connected with the formation of the lower part of the composite spine in the posterior line, and thus accompanies the turn inward of those muscles from their upper parts.

§ 130. The *fore-limbs* we suppose to be guided by the anterior part of the trunk, and by the neck, in the following manner. The edge of the shoulder blade, opposite to the arm socket, or more exactly speaking, that perpendicular to the prolongation of the central axis of the movements of the anterior limb* lies on the *posterior* true ribs, near the spine, so as to be guided by the propellers of their convex (§ 62), and thus follows their movements.

---

* This would not include the whole of the posterior part of the shoulder blade, some portion being under the guidance of the neck. Referring to the best engravings to which we have access, there is a difference in the proportion of the shoulder blade, which might be supposed to be guided by these ribs; the part under the influence of the neck being much larger in the lion than in the horse and giraffe.

That *end of the shoulder blade which carries the shoulder joint socket* lies on the anterior ground end of the true ribs, and is guided by the movements of this part, which amounts to *following the movements of* the front part of bearers of the convex (§ 62) and *the upper end of the sternum.**

The upper end of the sternum moves with the sterno-mastoids, and they, finally, move with the neck S.

At the formation of the left anterior winding-line, in both its sections, the propeller true ribs (on the rear portion of the convex) would so move the right shoulder blade that the right fore-foot will rest on its outer bearing; the establishment of the primary section will so alter the shape of the ribs as to bring the shoulder socket more against the bearers. The upper end of the sternum is carried somewhat toward the left.

At the formation of the right posterior line the upper end of the sternum is carried toward the left, at least relatively, by the movement of its posterior end to the right, as the left hind foot comes on to its inner bearing. This movement of the sternum brings the right fore-leg on its inner bearing by the movement of its shoulder socket.

---

* The clavicle which exists in some animals would compel this.

From this position the spring takes place, and the fore-foot, when lifted, continues presenting its inner bearing, not only because of the position in which it left the ground, but from its being presently again governed by the upper edge of its shoulder blade, which, on the change of curvature, lies at the rear of a *concave* of the true ribs. Thus it remains until the formation of the alternate winding-line which has discharged the old curvatures in the ophidian and composite S S, extends to the neck S, brings the (now free) right fore-leg in connection with the new posterior convex of the neck and causes it to present for the outer bearing.

Thus we should have in the trot for a fore-foot, while in appui, the outer bearing from the upper edge of the shoulder blade, while the anterior *line of its formation is in action. The inner bearing from the socket, while the posterior line of its formation is in action.*

But *a raised fore-foot,* having the posterior edge of its shoulder-blade on the rear of the new concave of the anterior C of the ophidian S *remains, presenting for the inner bearing until the* NECK *formation of the new anterior line puts it on the outer bearing,* by connecting its socket with the posterior part of the rear C of the neck S.

§ 131. Although the ophidian, composite, and bi-composite spines begin the development of the winding lines, separately, yet, once begun, each continues its development throughout; that is, until the anterior winding line, beginning in the ophidian S, has formed its bi-composite portion, the ophidian action constantly increases along with the others, and so also for the posterior winding line. Thus, the final action still centres between the upper and lower lung lobes (§ 99).

§ 132. It will be remembered (§§ 32, 33, 34) that to bring both C C of any S into a discharge forward, the cross-line must be displaced, and the posterior C rests, for its point of reaction, against the plane in which the anterior point of application moves. For this reason the ophidian posterior C thus rests against the neck-root plane; the posterior C of the rear half of the composite spine (its junction with the central S being formed) rests against the same point. The final point of reaction will be at the base of head, to which the points of all the spines will be transferred.

These posterior C C, their appuis (the hind-leg, &c.) being fixed, thrust their points of reaction forward, and (though the anterior C C discharge, so far as the spring is concerned, before them) their appui, *the hind-leg, precedes in leaving the ground,* the appui of the anterior C C, *the diagonal fore-leg,** the latter being in great part used only as a rest while collecting the spring and in lifting the anterior part of the body. It is discharged only when the whole composite S springs from the ground.

The posterior winding line prepares the discharge of both C C on the already formed anterior line, but its working is not completed until the posterior "point of application" has brought the posterior end of the cross-line to its own side, as the anterior point of application has already brought the anterior end. This cannot be done until the anterior end, which holds it in check (§ 32) is completely liberated, consequently not until after the liberation of the spring from the hind-foot, where *the direct drawing of the posterior point of application on the posterior cross-line end completes—first, the change of curvature in the anterior C, lifting up the fore-foot, and then, completing the change of curvature in the posterior C, puts down the*

---

* In our usual illustration, the parts would be the left anterior and right posterior points of application, the left hind and right fore legs, and the cross-line would be, at first, drawn completely over to the left side, the anterior end being drawn, and the posterior end held in check there by the left anterior winding line.

collateral hind-foot which is to form the new rear appui.*

§ 133. The ball and socket movement of the *digastrics*, and of the internal oblique muscles of the eyes, we should describe on the principle that a rolling outward of the muscle, beginning at its anterior part, represents the ball motion—a turning inward, commencing in the rear portion, the reciprocal socket motion (§ 39). The eyes remain steady in the central line of advance while, with movements similar to those of its diagonal eye muscle, each digastrics brings the composite spine into connection with the central movement. The eyes thus belong fundamentally to the ophidian (or snake) spine.

The part of both digastric and eye muscles anterior to the loop rolls outward with the secondary section of the anterior winding line, the part posterior to the loop with its primary section—then, to form the double twist, the rear portion rolls inwards with the primary section of the posterior winding line, the front portion inward with the secondary section. *These correspondences take place for the eye muscle during the ophidian spine action—*

---

* The newly grounded hind-foot should, in perfect action, come into the print of the fore-foot which gives place to it, unless the spring has carried all four feet forward.

for the digastric during the formation of the composite spine.

As the left anterior winding line enters the "bi-composite spine" formation, the revolution outward of the left digastric is completed, and its loop goes to the extreme rear limit of its rear movement.

At the formation of the *composite spine*, in the *right posterior winding line*, the action of the right digastric, turning outward from above, beginning with its posterior end, which really represents the socket movement, becomes transformed, (§ 39) as to effect, into the left digastric-turning inward, beginning at the rear.

§ 134. The movement of the *left hind-foot* in appui, following the left lower lung lobe, corresponds with the movement of the left (its collateral) digastric, so long as the composite spine is forming. But, so soon as the bi-composite spine forms and the *right fore-leg* comes into correspondence with the left (its diagonal) digastric, this leg *substitutes itself for the digastric in the connection with the left hind* if the movement penetrates the neck root joint.*

---

* There are some movements, notably, the walk, in which it seems to us, that the neck root joint shirks more or less of its full actions. (See § 146).

Thus the right eye, the left digastric and the right fore leg are brought into one direction, first by the anterior winding line, then by the posterior.

§ 135. The *sternum*, which is influenced by both formations, *turns—with* that of *the left anterior winding line at its upper left corner* when forming the bi-composite spine, so that the left sterno mastoid revolves outward in its lower portion, and allows the "displacement" of what would be the left anterior cross-line and of the neck S—at about one-third its length from the head. With that of *the right posterior line* the right side of the *lower and of the sterno* (since it represents the separating head-joint socket) is turned outward, but the *tension comes on the left side turning inward as the pressure* from the right socket movement came on to the left.

These movements of the sternum, in which its left lower end may be considered as representing the posterior end of the cross-line, and its right upper corner the anterior end, *give a full collateral traction just* as the full gathering takes place.

In consequence of the delay in the *full* discharge of the posterior C (§ 132), *the free* (right) *hind-leg does not come into connection with the right digastric, for its future appui, until the alternate anterior winding line is fully developed*, whereas *the free left fore-leg*, moving with the neck, *is in connection with it from the first formation.*

We have dwelt on these points so long, because, although they may seem slight, they are of great importance in "setting up" where the removal of each chronic displacement adds to the effect as the point approaches the final centre in a continually increasing ratio. In developing the action of the bit in the horse's mouth, the connections of the digastrics are also of fundamental importance.

§ 136. *For the* COMPOSITE SPINE *we may trace the anterior winding line — for instance, the left one — as beginning at the spine on the left side of the neck-root, passing around the body on the right side, so as to cross it in rear of the upper lumbar vertebra, and finishing in rear of the left (collateral) hip-socket.*

The right posterior winding line, as beginning at the right hip-socket in front, passing to the lower end of the sternum, thence around the left side of the body, and under the right (collateral) shoulder-blade.

§ 137. *For the* BI-COMPOSITE SPINE, with the *head-joint* as the seat of the "points of application," *the condyles (working down into the lungs) representing the "anterior points," the sockets the "posterior," the* shoulder joints combine the character of both sets of limbs, and the lower jaw is the substitute ground.

§ 138. *The eyes*, if we trace the optic nerves (which we have attempted to do further on), may be considered to *combine both rear and front appuis as the foot ends of the whole system*, first for the ophidian spine, and successively for all the others, as they form. *For the trot they first assume their relation to the ophidian spine, and then, by the action of the neck, the bi-composite relation is superadded.* For a variation from this gait the lower jaw, head-joint, and shoulders, may first act, and then on the eyes the composite spine may be brought in.

§ 138. The pelvis and the shoulder-blades act with both C C. *The third S is so rigid in its composition that it is difficult to separate the movements of its C C, except by their effects. At the formation of the bi-composite spine the whole third S is merged in the rear C of the ophidian, as is the whole neck S in its anterior C, bringing the cross-line centre of this spine between the lung lobes.*

§ 139. The lower jaw is the ground of junction between the composite and bi-composite spines, and on it the re-actions from the real ground, as well as from all the substitute grounds (§§ 81, 82, 125) come in bearing. These are essentially *collateral*.

In the same way, the shoulders receive all the *diagonal* actions.

In giving the details of the trot, which we are about to do, we have followed out each section of the winding lines as if it were developed in each of the spines — ophidian, composite, and bi-composite — in succession, before the succeeding section began in the ophidian spine. This has been done to avoid the confusion from intermixing them; but it is obvious that the ball and socket motion of the neck cannot succeed the anterior line of the ophidian spine *smoothly* until the primary section of the latter is partially formed, since this latter moves the neck-root joint. Hence, the ophidian and composite S S of the trot are always in advance of the neck S — that is to say, both their secondary and primary winding-line sections are in formation before the secondary of the neck begins.

§ 140. It may be mentioned that we shall consider THE TROT as *formed by commencing with the ophidian S*, and through the continuance of the action of this S developing the other S S, and, finally, the bi-composite spine; THE PACE as *commencing with the neck S*, which, by its development, causes the formation,—first, of the third S, through the sternum, and then of the ophidian; THE WALK, as *commencing with the ophidian, but not extending regularly*, the neck-root working being only partially performed; in fact, as being a trot with the hind-legs and a pace with the fore.

As to the gaits of double action we shall consider *the* SQUARE JUMP, which may form *a species of gallop*, to be a *double trot;* the " FULL RUN " to be *a double pace;* THE CANTER to be *a mixture of the trot and pace*, which is *finished in each spring by double action*.

DETAILS OF THE TROT.

§ 141. Let us, as usual, suppose the left anterior winding-line to be the leading one, we will mark the developments by the following symbols:

A for the original anterior winding-line.
B " " " posterior "
A′ " " alternate anterior "
a " " secondary section of the anterior line.
a′ " " primary " " " "
b " " secondary " " posterior "
b′ " " primary " " " "
a^1 and b^1 for the ophidian spine.
a^2 " b^2 " " composite "
a^3 " b^3 " " bi-composite "

§ 142. A a¹. Formation of the secondary section of the anterior winding-line in the ophidian spine. The "point of application" being on a cross-plane section, somewhere at the root of the neck, the ophidian left anterior winding-line develops from it. This development causes the left false and right true ribs to form on convexes of the spine, which work against the sternum, and at the same time expand the left posterior, and, although less fully, the right anterior lung lobes, at the expense of their reciprocals, the collateral left anterior and right posterior lobes.* The consequent changes in the centre of gravity bring the *left hind-foot* and, in a less degree, the *right fore-foot* on their *outward* bearings, giving them also a greater share of the weight of the body, and by so much, relieving the right hind and left fore-feet. With the movement of the false ribs the left side of the diaphragm is stretched.

The *left fore-foot* is turned in by the commencement of an inward facing of its socket (§ 130), following through the upper part of the shoulder-

---

* The anterior portion of the eye muscle rolls outward with the *filling* of its diagonal posterior lobe, and inward with the *exhausting* of its collateral posterior lobe. Hence, since it cannot roll two ways at once, the two posterior lobes cannot be reciprocals as to filling and exhausting.

blade the shape of the rear true ribs of its side. The *right hind-foot* is turned in by the facing of its socket.

The right eye becomes the centre of direction, the muscular portion in front of the loop rolling outward, and thus the ophidian spine preserves, in the eye, its relation to the snake's motion.

a'^1. Formation of the primary section—Upper end of sternum carried somewhat to left, brings bearing of right shoulder blade on the ribs, nearer to its socket part—*Displaces ophidian cross line* to left—Rear portion of right eye muscle rolls outward, loop slipping to its rear limit—Enlarges right upper lung lobe.

a^2. This movement combines the third S with the ophidian spine, and thus forms the rear portion of the composite spine for the anterior winding-line. As the internal relations of the composite spine belong to the ophidian S, and the external to the neck S, the anterior "point of application" for a^2 a'^2 must vary between the "point in a cross plane" spoken of under a^1 and the head condyles—Turns the *left* hind-foot still more on its outer bearing—Fills the rear C of the right lung tip (in filling this, revolves to the left and backward)—Tightens connection between left side of pubis and left side of posterior end of sternum, in the outward bearing, in consequence of the turning of the lower of the third S. Turns the upper part of left diaphragm pillar outward, and still further expands the left lower lung lobe.

a'^2 Formation of anterior part of composite spine in anterior winding line—Rolls left anterior corner of sternum outward and backward, and causes lower part of left sterno-mastoid muscle to roll in same direction—completes outer turning of left diaphragm-pillar, and begins that of left psoas. The turning of the left upper corner of the sternum causes a *partial movement in the neck root joint*, in its character of a ball and socket—Right fore-leg shoulder socket brought close to ribs.

a^3. Continued revolution of left anterior ophidian point of application further moves neck root joint, so that left head condyle turns inward by the rear on its posterior end. The *left digastric* rolls outward in its anterior portion, the loop slipping backward. The *left* (free) *fore-foot* turns outward.*
The left psoas completes its outward turn, thus forming the *rear portion of the bi-composite spine in the anterior line.*

---

*When the trot is fully inaugurated, the anterior line which we are now describing, is the *alternate* line in relation to the previous spring.

a'^3. Continued action of the ophidian line works still further on neck root joint; the left head condyle, in consequence, rotates forward, so as to pass its interior side against the interior side of its socket, resisting the movement toward the left coming from the right fore-leg, and carrying its own point of pressure forward; thus corresponding to a^1 § 37. The rear portion of the *left digastric* rolls outward, the loop moving to its utmost rear limit. The left fore-foot is carried forward.

The movements a^3 a'^3 of the left head condyle are made on the rear C of the right lung tip, and the *bicomposite spine* for the "anterior line" is completed in the formation of its front part by (in this case left) side of the lower jaw setting, in its socket, so as *to rest* on the inner part of its condyle. This bearing, and that on the outer side, are, respectively, analogically the same as the outer and inner bearing of the ribs and feet; we shall, therefore, name them the *analogical outer* and inner *bearings* of the lower jaw.

§ 143. B. It is a little difficult to describe the action of the posterior winding-line as it covers the precedently established diagonal anterior line. Its secondary section working in the anterior C, is there opposed by the primary section of the anterior line holding the cross line displaced, and the first effects are, on this account, not those of a working in this C, but *by re-action* of a working in the posterior C. Here, beginning at the rear end, it double twists the propeller false ribs, bringing them on their inner bearing.

With the establishment of its primary section the posterior winding-line would move the posterior cross-line end, but this is held fast, displaced to the left, until the anterior line is discharged in its primary section by the action of the primary section of the "alternate line," when it displaces its appropriate end of the alternate cross-line.

In consequence of this delay the propeller true ribs are, next after the double twisting of the propeller false ribs, brought on to their inner bearings, and even the bearer true and false ribs are brought on to their outer bearings (§ 53) while the displacement is still maintained.

At the *discharge*, which begins with the formation of the secondary section of the "alternate anterior winding-line," the original posterior C is discharged, *virtually*, but it is only when the establishment of the primary section of this "alternate line" takes place, and discharges the original anterior C, releasing the original anterior cross-line end as it draws that end of the alternate cross line toward the right, and releasing also the original posterior cross-line end from its constrained position to the left, that this virtual discharge is carried out and the alternate false rib propellers are grounded.

The analogy of the snake's action holding good for the three spines and their appuis, the right hind-foot will be set down when $A'\ a'^2$ has released the cross-line end of the composite spine at the juncture of the lumbar and ophidian vertebræ.

As has been so frequently said (§ 39, etc.), all the actions of the posterior lines are "replacement" movements in the side of the C which they affect. Thus the inward rolling of the left digastric is really consequent on an *inoperative* outward rolling of the right digastric, in the same way as the pressure of the left head socket comes from the drawing away of the right one.

b'. Posterior right winding-line. Effects in posterior C of the re-action from its secondary section as checked in anterior C, brings the left propeller false ribs on to their inner bearings, beginning at the rear, and produces an effect in the further stretching of the left side of the diaphragm by "counter-action" in its fibres. Begins a rolling motion of the right eye muscle inward, in its rear portion corresponding with the motions, from re-action in the rear C of the spine. Produces some inward bearing of the left hind-foot from the change of centre of gravity.

b'. Direct action in secondary section. Right true rib propellers come on inner bearing, and bearers on outer bearing. Right eye muscle rolls inward, by its anterior part, loop still not moving.

b^2. Re-actionary formation of rear half of composite spine for the posterior winding-line. Brings the *left hind*-foot on to its *inner bearing*. Rolls the left pillar of the diaphragm inward by its lower part. Turns (by change of socket facing) the right (free) hind-foot for outer bearing.

b^2. Direct formation of anterior part of the composite spine, for right posterior winding line, secondary section. Lower C of right lung tip fills, rolling inward. *Right fore-foot* comes on inner bearing by movement of socket from continued passage of the upper end of sternum to the left. Turns the posterior end of sternum, left side, inwards by the front, so far as the displaced position of the cross-line end will allow it.

Partial movement in neck-root joint crossing

that of a^1. Left diaphragm pillar rolls inward at its upper part. Left psoas rolls in by its lower part.

b'^3. Reactionary neck S. Since the action is spreading upwards from the neck-root joint, the movement of the digastric will somewhat precede that of the head socket. Left digastric rolls inward, beginning with its anterior part, but loop does not move. Left head socket rotates forward so as to pass its exterior side against the exterior side of the condyle, still resisting the general pressure to the left, which is now exerted through the *condyle*.

Action through the sternum forms *rear part of bi-composite spine for the posterior winding line* connecting *left* (fixed) *hind-leg* with the left digastric, and by the diagonal part of its traction bringing the *right* (free) *hind-leg* forward. The left psoas rolls inward by its upper part to form in the bi-composite spine.

b^3. Direct pressures and counter-pressures gather at the anterior point of the left head-joint socket. Left digastric rolls inward by its posterior portion. Left side of lower jaw comes as firmly on to its "analogical inner" bearing (§ 142) as it can, until the formation of b'^3.

Movement on lower C of right lung tip finishing

the motion of the neck-root joint *completes bi-composite spine anteriorly for the right posterior winding line*, excepting the formation throughout of b', b'^2, b'^3, its primary sections, which await the release of the posterior cross-line end from its position to the left, on the formation of the alternate anterior line.

Right anterior lung lobe and lung tip straighten into one. The sternal action would connect the right fore-leg with the right digastric, but for the same reason (§ 39) that the action of the right head-socket only gives pressure on the left; this only causes accumulated action in the left digastric.

DISCHARGE. By the formation of the right *(alternate)* anterior line.

A' a^1 a'^1. Gives *final* discharge to the ophidian S so soon as a'^1 draws the anterior end of the alternate ophidian cross-line to the right, thus releasing the end of the old line.

A' a'^2, moving the anterior end of the alternate composite spine cross-line, after *fully* discharging the left hind-foot by releasing the "reactionary," and the right fore-foot by releasing the "direct" action of the secondary section of the old posterior line, sets down the free right hind-foot by the release of the old posterior cross-line end of the composite spine. A' a'^2 likewise frees the left digastric loop after allowing the now, for the first time, possible action of b' b'^2 of the old line. This loop passes instantly forward to its utmost front limit, and at the same time allows the left jaw to set firmly on its "analogical inner bearing" (§ 142)* given up to the drawing of its temporal muscle.

The new appuis being now set down on the outward bearing, and the old ones being raised as they left the ground with the inner bearing presented.— A' a^3 and a'^3 form, in some horses, with great rapidity, and perhaps a^3 before reaching the ground, but in others with a slight delay.† These turn the now raised (here right) fore-foot for its outer bearing and bring it forward; then, when B' is formed, the left (now free) hind-foot is turned out by B' b'^2, and b^2, is brought forward by B' b'^3 and b^3, and so on.

Whether, in the landing, the new hind or the new fore-foot reaches the ground first, depends on the degree of gathering (§ 64.)

§ 144. THE PACE. We include under this name

---

* We suppose the sudden pull on the rein collateral with the propelling hind-foot, at the moment of the thrust, to be owing to this sudden setting of the lower jaw. It is well known that drawing this rein *just* as the foot has left the ground, most powerfully checks progression.

† Where this delay is very marked, the lifted fore-foot not being presented for its outer bearing, retains the inner, which is called "clishing."

all those methods of locomotion in which quadrupeds use *two legs* of the *same side* as *appuis for a step*.

In the trot, the movement was begun with the ophidian S acting with the eye diagonal to the rear appui, and, extending to the third S, and finally to the neck S, concluded by this S acting with the digastric collateral with the rear appui. We suppose *the pace* to be begun with the neck S and the collateral digastric, to extend to the third S, and finally to be concluded by the ophidian S and the diagonal eye.

There was what might be called an "ophidian action" of the head articulation coming from the spine in the eye movement of the trot, but we have now the thorough action of the separate condyles; this ended the trot, but begins *the pace*, and the sterno-mastoids bring the tractions first to the sternum. The *sternal tractions* are (§1 18) *collateral*, and *hence* the *point of appui* will be *two feet of the same side*, instead of as in the trot, diagonal.

In our theory of the trot, the back bone is drawn over the sternum and the final stress of the discharge brought on to the front of the latter. In that of *the pace*, the sternum is first drawn forward and the final stress of the discharge brought towards the root of the neck (or perhaps the withers).

It is apparent that, for animals which carry great weights in their mouths, as the lion with its prey, the pace—since it *may* commence with the lower jaw as its initiative point—would be the most advantageous. The single-action locomotion of the lion, &c., is, we believe, of this nature; and it is noticed in such animals, that the lower jaw articulation is a simple hinge, without any lateral movement,* so that the body must rather move about the lower jaw, than the lower jaw accommodate itself to the movements of the body.

These movement of the pace, however, would more *properly* begin with a head-condyle and its thorough action. It is in this way we suppose them to be performed by the Horse and the Giraffe.

The digastric loop may be released either with a lateral movement of the lower jaw as at A' a'^2, or by an extended lateral movement of the diaphragm as the base of the lungs, or by a lateral movement of the head as the basis from which the digastrics hold; which last again may be supplied by a still freer lateral movement of the jaw. In the trot of

---

* Possibly the fact that, from the absence of the bony partition between them, the temporal muscle presses directly upon the eyeball in these animals, may have something to do with so connecting the digastric and the internal oblique of the eye that the lateral motion of the jaw is less needed.

the horse the lateral movement of the lower jaw, as it leaves its outer bearing, supplies whatever is wanting in the similar movement of the diaphragm. In the lion, &c., the long flexible body gives sufficient movement to the diaphragm for replacing entirely the movement of the jaw. But in the Giraffe there must be an extended jaw movement, to replace the want of proportional lateral movement of the diaphragm caused by the long neck and comparatively short body.

When *moving with the neck, the fore-limbs are related to the neck S,** *as the hind-limbs are to the third S,* and thus the limb on the side of the rear convex of the neck will receive an outer bearing turn, the one on the side of a rear concave an inward bearing turn. This relation will be more marked when the neck S acts first as in the pace.

The *third S is for the rear what the neck S is for the front* (excepting so far as the latter is more concerned in forming the bi-composite spine). After (for the pace) receiving its development in the *anterior* line by the extension of the neck S, and when *this* developement has been extended to the ophidian S, the *third S commences the formation of the* POSTERIOR *winding line.* From the beginning, its

---

* We would recall the fact of the fore-limbs being head-limbs.

development is in close counter-action with the neck S and the latter after the formation of the ophidian part, developes the bi-composite spine for the line by the final thorough movement of the condyles.

§ 145. Referring to § 141 for the symbols used, we should *form the* PACE by the following combination :

A a^s. The left head condyle revolving at its posterior end, independently of any preceding ophidian movement, throws the *weight on* to the *left fore-leg* by its *neck* connection ; and the left sterno-mastoid muscle turning outward by the movement imparted to the left upper corner of the sternum, through the formation of the left convex in the posterior C of the neck S, turns the *left fore-foot on its outer bearing*. This working is, in the horse, directly assisted by the Levator humeri muscle (§ 86, note*). There will also be a certain effect in throwing the *left hind-foot on its outer bearing*.

a'^3. The left-head condyle passes its interior side against the interior of its socket. If the left fore-foot were free this would carry it forward, but this foot being fixed, the traction of the sterno-

---

* Were the foot in air, the drawing of this muscle would bend the leg ; being in appui, the straightened leg draws on the neck.

mastoid becomes diagonal through the sternum, and brings the (free) *right hind foot somewhat forward* with its *inner bearing* presented.

a^2. The rear part of the composite spine, beginning with the third S, forms, by induction, from the neck. The *left hind-foot* is brought *more fully on* to its *outer bearing*.

a'^2. *Right fore-leg bent* and its *foot* presented for *outer bearing* by movement of sternum to the left, changing the socket facing as the neck root joint moves.

a^1. Completion of outer bearing of *left hind-foot*.

a'^1. Completion of action of *right fore-foot*, as in a'^2 but its movement now derived from the rear true ribs.

B b'^3. Begins with the third S, but, as before said, so in counter-action with the neck S that they move almost together. Throws *left hind-foot* on *inner bearing*. Turns *right hind-foot* on *outer bearing* and *advances* it.

b^3. Brings *left fore-foot* on its *inner bearing*.

b'^2. *Right hind-foot forward outer bearing*.

b^2. Right fore-foot inner bearing.

b'^1. Increase of b'^2.

b^1. Increase of b^2.

A' a³. *Discharge. Turns right fore-foot* on *outer bearing.*

— a'³. Brings right fore-foot forward.

A' a³ *discharges* the *left hind-foot* in its bi-composite connection, and — a'³ *the left fore-foot* in the same. These discharges, of course, work on the sternum, and are followed by the composite and ophidian spines; but, from the circumstance of the altered position of the appuis of the alternate anterior line, the collateral fore and hind-leg are not much separated after leaving the ground.

§ 146. One difficulty in describing all the gaits lies in the difference between the action of the anterior line when first throwing the body into position, and its action when as "alternate anterior" it discharges the gathered three springs of the old lines. In imagining the real working of this line, when the gait is fully inaugurated, we must bear in mind that, then, the first office of the anterior line is to discharge the old spring, after which it forms the basis for a new one.

It will be noticed that a marked distinction between the trot and the pace consists in the "alternate neck winding line" A' a³ a'³ forming itself in the trot, *after* the spring and landing, but in the pace, *before* either.

In the trot the discharge of the C C succeeded each other with a marked interval; that is, the rear C was *virtually* discharged and the alternate anterior line for this part formed, before the anterior C was discharged and its portion of the alternate anterior line was formed, and, in retrogression, vice versa.

But, since the pace begins at what was the conclusion of the trot, the two C C are discharged more together, and the discharging alternate anterior line, having awaited their discharge, then forms more as a whole; that is, *in the trot* the left hind-leg virtually discharges and the right posterior lung lobe fills, before the right fore-leg discharges and the left anterior lung lobe fills; whereas, *in the pace* the left hind and right fore-leg discharge (this last not in appui), then the right lower and left upper lung lobes fill.

The above described difference will make a prominent point in "setting up," causing a distinction between forcing the trotting and the pacing movements to the "third result" (§ 76).

The Walk is described by von Oeynhausen as the gait in which each foot moves at separate intervals, and so that, while in moving forward each fore-foot is succeeded by its diagonal hind-foot, each hind-foot is succeeded by its collateral fore-foot. Thus, supposing (1) the right hind-foot to move to the front, then (2) the right fore-foot next moves, and, for an instant, the horse is resting on the two left feet; (3) the right hind-foot is put down, the horse is on three feet; (4) the right fore-foot is put down, the horse is on four feet; (5, 6, 7 and 8), the same is repeated, with the opposite side leading.

§ 147. We should construct the *walk* from the elements of locomotion A, B, A' given in § 141, by supposing the action of the *primary* sections suspended in all the anterior lines, and the *secondary* sections of the posterior lines to be carried only far enough to secure the necessary reactions in the posterior C C C. In this way, the through ball and socket action of the neck-root joint is passed over, and the fore-feet given up almost entirely to the neck action; the hind-feet to the body action. From this gait the extension of the neck S lines, when they are developing, will produce the pace; the extension of the ophidian lines, the trot. The passage of the outer and inner edges of the condyles will be very imperfectly made, and the passage from the anterior to the posterior condyle ends will thus take place with something of a jerk, which gives a "nodding motion" at each step, particularly marked in good walkers.

A. a^1. Puts *weight on left hind-foot.*

a^2. Increases this joining the ophidian and third S S.

a^3. Puts the *weight on left fore-foot. Raises right fore.*

B. b'^1 and b'^2. Brings *left hind-foot* on its *inner bearing;* turns *right hind-foot* for *outer bearing* and brings it *forward.*

b'^3. Puts *left fore-foot on inner bearing.*

A'. a^1. Releases $a^1 b'^1$.

a^2. Releases $a^2 b^2$ at junction of central and third S. *Puts down right hind-foot.*

a^3. Carries *forward and puts down right fore.*

Thus we have A' a^2 right hind foot down. A'^3 right fore foot down. We are inclined to think that in the *full progress* of the step a fore-foot *leaves* the ground as the opposite fore-foot is put down, thus slightly differing from the description of von Oeynhausen; but as this fore-foot, after rising, awaits the movement of the collateral hind-foot, raised and carried forward by the posterior line, the difference is not important.

§ 148. *Retrogression* may be brought about, after the formation of the diagonal oppositions which prepare for the spring forward, by refusing the discharge and continuing to develope the original lines of gathering. While the alternate anterior line is kept from forming (§70 snake's motion), the head-joint, both as to its ophidian and separate condyle (or limb) motion, will change to the *alternate* bearings, and the effect of this will be—*first*, to change the appuis to the alternate legs; and *second*, on these *concave* appuis to discharge the old lines and form the new ones.

This delaying the spring until all the concaves have *actually* changed to convexes, and thus making that change on the ground entirely, instead of finishing it in air, again brings the head condyles, in advance, into their alternate position, and puts the concaves in appui for the next step.

We think that *five points of difference* from progression must be produced by that accumulation of action at the head-joint which, in retrogression, at each step keeps the balance of gravity on the concaves. (1.) The filling of the alternating lung lobes must be in advance of that in progression. (2.) The passage of the convexes over the concaves (§ 60) must be reduced to a minimum, if, indeed, the contrary action do not have place. (3.) The thrust of the feet of appui, depending now on the actual change of the spinal convexities (§ 61), the movement of the diagonal feet of appui, in the backward trot, must be more nearly synchronous than in that of progression. (4.) The action of the lower jaw must, in proper backing, be that of yielding, instead of forcing, at the moment of the spring, the alternating joint leading the working one.

(5.) *In backing (§ 70) the feet leave the ground by the outer bearings and take it again by the inner bearing*, exactly the reverse of progression.

§ 149. *The Halt.* In the "the third result" of forcing the winding lines in the snake (§ 71), we endeavoured to describe that of bringing on the formation of one line by *induction* from the continued forcing of the other, after the *allowed* counteractions had been already formed; also to show that so doing would bring on an equalization of gathering in the two sides, resulting finally in the "superimposition of twists," a formation in which all four winding lines are developed at the same time, and consequently all *lateral* curvatures suppressed.

As the actions of halting are identical with those on which our theory of "setting up" is founded, we shall postpone a part of the details to the sections in which that subject is treated.

In retrogression both the anterior and posterior lines were *separately* formed, but if one line be forced and the other form by induction, then, from the continuance of this forcing, we shall have as the result, equalization of gathering on the two sides, and thus a position from which any gait may be initiated by a redistribution of the tractions.

Any portion in the course of the winding line may be selected for commencing the forcing, and the equalization will then begin at the point which would have been formed by that moment of action in the original formation of the line.

We shall, however, for the present, consider the forcing as beginning in each S, at a point of application.

When a horse lands from any spring, he may be *halted* in *three ways.** *First* on the *alternate* anterior line of the *ophidian* spine—this line having being brought into superadded action by the pressure of the legs or by the spur, while the bit checks the extension of the line to a^3 in the neck S until this is formed as part of the adjustment in equalizing the sides from forcing its ophidian portion. In this case the horse will be halted on the outer bearings of the feet, and then come on to the inner bearings; i. e., he will be halted in progression, probably the most advantageous method.

Second.—On the alternate posterior line of the ophidian S. In this case the bit is drawn as the posterior line begins to form, after landing, and the spur is applied after the bit. The action of the neck S in the anterior line ($a^3 a'^3$) having preceded it, a fore-foot will have been raised. In this case the horse will be halted on the inner bearings, beginning with the hind leg of appui, and then come on to the outer bearings, i. e., he will be halted in retrogression. This is not a false halt, and is we believe the one generally used by the Arabs, and perhaps that used by the animal in a state of nature. *Third.*—The rein may be drawn after landing, before the action ($a^3 a'^3$) of the neck is completed. As the neck inaugurates a pacing movement, the halt will then be made in a pace ingrafted on the trot, causing a very awkward equalization of gathering, and this we suppose to be the "halt on the shoulders" so much deprecated by all horsemen.

§ 150. *Double-action Motion,* from position of

---

*We refer to the trot in exemplifying the subject.

"Superimposition of twists" (§ 71); i. e., spring by two rear and two front appuis.

§ 151. The *Double-action Trot* or Square-Jump must differ very considerably from the single action in the movement of the feet, since one fore-foot can no longer be thrown forward by the neck movement, $A\ a^3$, a'^3 (§ 143), while the other fore-foot acts in appui; on the contrary, the fore-feet when thrown forward must depend for support on the hind feet, or on an impetus derived from the spring of the hind feet. And, also, since *all the legs* must be *affected*, before the spring, *by both the body and neck gatherings*.

§ 152. We shall be obliged to add to the symbols, § 141, a character representing the *right* anterior and *left* posterior winding lines which can no longer be regarded as alternate, by giving an Italic letter to these last, thus $A\ a^1\ a''^1$, A', and so on, for the right anterior lines, $B\ b'^1\ b^1\ B'$, for the left posterior.

§ 153. In the double trot the anterior winding lines, left and right, acting together throughout their secondary and primary sections, in the *ophidian S*, throw, by change of centres of gravity, the weight on the outer bearings of both hind and both fore-feet. Cause both internal oblique eye muscles to roll outward, &c., &c. A A a^1 a^1 a'^2 a'^2.

a^2 a^2. Form the rear of the composite spine. Cause the back, from the junction of the third and ophidian S S to the tail, to begin forming a convex in the medium perpendicular plane*. Increase the *outward bearing* of *both hind-feet*.

a'^2 a'^2. Forms the anterior part of the composite spine. Moves the neck-root joint as ball and socket for these lines. Increases *outward bearing* of *both fore-feet*.

a^3 a^3. Both digastrics roll outward. Both loops drawn back. Muscles connecting pubis and sternum tightened. Bicomposite spine formed in rear for anterior line.

a'^3 a'^3. Fore-feet fully on outward bearing. Digastric loops to rear limits. Motion of condyles and motion in neck-root joint completed for anterior lines. Bicomposite spine completed for the same. Both lower jaw articulations set on their analogical outer bearing.

B B. In the ophidian and composite spines these

---

* The perpendicular convexities are properly two for each S S S, but, as the bicomposite spine forms, the various portions are so reduced as to form two for the whole body, one the neck and part of the ophidian spine, the other, part of the ophidian spine and the third S. Between the two the centre of the ophidian spine lies in concave.

lines bring the hind and fore-feet on their inner bearings. In the bicomposite spine this bearing is increased, and the whole body brought into one gathering with two convexes, upward, in the median perpendicular plane; i. e., one consisting of the neck and part of the ophidian spine, the other of part of the ophidian spine and the third S; the centre of the ophidian spine sinking as a concave between them.*

A' A'. Release the gathered springs of all the spines successively before leaving the ground, and beginning with the ophidian spine. The hind-feet are thus released first, and the fore-feet immediately after them. The release of the fore-feet by the alternate neck lines—which in the single action trot would have been accomplished after the landing—is in the double action accompanied by a sudden and rapid carrying of them forward, at the time when the head condyles change their bearing.

The feet come down gathered on the anterior lines, and immediately form on B' B', and so on.

§ 154. At one moment of the "double trot" the horse is much extended, the fore-feet being

---

* Here again we have the centre between the lung lobes as the "centre of force" for the whole body (§ 99).

stretched forward and the hind-feet just drawing up from the thrust backward. The whole action is so violent that it is unfitted for more than a few bounds.

The landing may be made on the hind or the fore-feet, according to the distribution and force of the gathering.

As the animal is in "double superimposition of twists" at each gathering, there will be no equalization of the sides at the *Halt*, which now can be made only on the alternate posterior line, and, by a violent and disturbing effort on the neck action.

§ 155. The *Double Pace* or *Full Run* ("Carriere"). This gait will bear the same relation to the single action pace that double action trot bears to single action. The *movement*, however, will differ widely from that of the trot, inasmuch as the whole bicomposite spine is first discharged by the sternum, and the hind-legs follow so closely on to the fore that they separate but little either in leaving the ground or on landing. As in the pace, the head condyles change their bearings while the appuis are still on the ground (in the double trot they change in air). The push at the bit should also be different, since the lower jaw articulations preceding the digastrics in their movements, it will occur before the *hind-feet* leave the ground instead of, as in the trot, just as they have done so.*

§ 156. The *Canter*† or *Gallop*. There remains still this other perfect gait, which is the *usual* method of locomotion in double action.

If we consider the double trot, in reference to the alternate expanding and contracting of the lung lobes, it will be seen that, in the gatherings, there is a very forcible expansion of them, in which state they are retained to await the movement of the breast-bone (or " substitute ground"). This is evidently a laborious action, as may be observed when putting a horse through the movement. On the other hand, although the double *pace* is easier, because the extreme tension comes on only at the moment when the completion of the final diagonal actions enables the lungs readily to relieve themselves at the spring, and makes this gait the one for the highest speed. Yet, since it requires a complete leaving of the ground by all four appuis at the same moment, and gives no intermediate instant of rest on two appuis, it calls for a great expenditure of force.

---

* It must be necessary for the lion, &c., when carrying a weight in the mouth, that the condyles and lower jaw should take their alternate bearings before the body leaves the ground.

† We shall use the term canter as more definite than gallop.

These objections existing for any ordinary movement, either in the double trot or double pace, a combination of the two is adopted as the usual method of locomotion with double action.

The *succession of the feet* in the canter, according to von Oeynhausen, is in *leaving the ground.* (1.) a' hind-foot; (2.) the collateral fore and opposite hind-foot, so closely in succession as to be almost synchronous; (3.) the diagonal fore-foot. In *coming to the ground* the order is the same, excepting that *when the weight is well on the haunches,* the hind foot of No. 2 anticipates its diagonal fore-foot, and in this latter case there will be *four* "beats," since each foot comes separately down, whereas in the former only three can be separated by the ear.

It is evident, we think, that von Oeynhausen considers that in No. 2 the collateral fore-foot anticipates the opposite hind-foot in leaving the ground, and that (adopting our usual illustration), we have—(1.) The left hind-foot; (2.) the left fore-foot; (3.) the right hind-foot; (4.) the right fore-foot. This would give the same relative "succession" as in the walk, the likeness of which to the canter he thus particularly points out: "The similarity between the *succession of the*

feet in the *walk* and in the *gallop* will not have escaped the observation of the reader.   *   *   *
It is also a frequently repeated remark, and one confirmed by experience, that the goodness or faultiness of the walk and gallop are nearly related to each other."

§ 157. We should explain von Oeynhausen's description of the canter, in accord with our theory of tractions, in the following manner:

We will suppose the canter to be "to the right," which may be best illustrated by assuming that the horse is moving around a circle of which the centre is to his right. The left legs are then the outer legs, the right legs the inner ones, and if the horse be cantering properly he "leads" with his right (inner) fore-leg.

In the "succession" of the walk—right hind, right fore; left hind, left fore. It will be noticed that the two legs of the right side successively pass the two of the left side, then those of the left the two of the right, and so on. This von Oeynhausen makes the basis of his description of the canter, and we shall use it for the same purpose.

One may imagine a gait in which the movements of the walk are performed with a double action.

'This would allow of no ball and socket action at the neck root joint, but would give a double trot for the body limbs and a double pace for the head limbs. Now, if in such double action a certain amount of *lateral* curvature—say convex to the left in the rear C C, to the right in the front C C— be allowed in the S S, and this curvature be constantly maintained, we shall have the double action trot of the hind-legs and the double action pace of the fore-legs, modified by a *moment* of single action for the legs of either side, as the winding-lines form, and in which the right-left counteractions will not at all enter the neck root joint, thus leaving the left fore-leg entirely to the neck action, as in the walk, and the left-right counteractions will enter to only a certain distance, as the partial actions of the left head-condyle and right head joint socket carry it. That is, the single actions of the body and of the neck, by which the gatherings of the left hind and left fore-legs begin, are, at their ending, merged in double action with the beginning of the gatherings for the right side legs, and the right side gatherings again *end* in single action.

The whole action of the right hind-leg is, when the rear C C are maintained convex to the left, made less forcible by reason of the left anterior crossline end being kept from moving to the right, and the movement of the left fore-leg is restricted, the anterior C C being maintained convex to the right, because the true ribs of the left side are kept from passing the sternum fully over to its support.

These C C being thus constantly kept partially convex, let us suppose that (1) the horse has in the walk put down the *left hind-leg* to gather on the anterior winding lines $a^1 a^2$, and that the gathering thus made is maintained, but kept suppressed. (2) He gathers on the left condyle for the *left fore-leg*, and this is also kept in abeyance. (3) and (4) He gathers in the same way for the right side. The gatherings for the two sides will *coalesce* at the ending of the one and the beginning of the other.

Let the posterior lines be formed in the same way for both sides.

There are now, as it may be said, "*latent*" four movements of the walk, and if, when the *alternate* anterior lines are formed to discharge them, the discharge be restrained, so that all of them spring nearly at once, the steps, so far as passing the feet is concerned, will take place *in air*.

The order of the spring will be :—(1) The left hindfoot (rear action exaggerated). (2) The left fore-foot (forward action limited). (3) The right hind (rear action limited).* (4) The right fore (forward action exaggerated) while in air, *first* the left hind and left fore-feet will successively pass those of the right side *as the horse rises*, fulfilling their steps; *second*, the right hind and right fore will repass those of the left side *as the horse descends*, fulfilling their steps. All the feet will then come down in the same order, and to the same positions as at the start.

In this gait, by reason of the permanent bendings of the body and neck S S, the left stifle joint and the point of the left shoulder are held always in rear of those of the right side,† and thus, although each foot passes, the *whole* leg does not.

If, during the canter, the horse extend the neck-root action from the neck backward, he launches into the " full run," if he extend it from the ophidian S forward he takes the *" double trot."*

Von Oeynhausen is inclined to explain the "*only two beats*" which are heard and felt in the *full run* as coming, not from the two hind-feet giving one sound and the two fore-feet another, but from the

---

* As before stated, the times of 2 and 3 are so close together in descending that the ear cannot distinguish them, and also if the haunches be weighted, 3 (the foot not the leg) precedes 2.

† Von Oeynhausen lays much stress on this.

left hind and left fore-feet giving one sound, and the right hind and right fore another. This may be so, and consequently, perhaps the double action never entirely lose a slight one-sided element, but this would seem to take from its perfection, and we should rather believe that the two sounds come from the slight difference in the times of landing of the two front and the two hind feet.

§158. The "*Disunited gallop.*" In this movement the shoulder of, say the left side, is in advance, while the stifle-joint of the *same side* is in rear. On our theory this must be explained by supposing the horse to introduce the wrong condyle into action with the ophidian movement.

§159. We shall leave the other actions of the horse for the subject of riding, since their description is so connected with that of the "aids" used by the rider for producing them, as to render it difficult to separate the one from the other.

PART V.

SETTING-UP.

Before rehearsing the chief points which are concerned in "setting-up," we will allude to some general principles.

It will be noticed that the alternate head ball motion, in gathering the alternate anterior line, coincides in direction with the old motion of the underlying socket, in gathering the old posterior line. Now this alternate ball motion absorbs the old socket gathering, and thus transforms it into the alternate anterior gathering, at the same time that it releases the old anterior gathering, which was held under control by the now transferred and metamorphosed posterior winding line.

We say "metamorphosed *winding line*" because the same thing takes place at all the articulations, and thus the (for instance) right posterior winding line becomes the right anterior winding line, the concave to the right of the lower C ascending to the upper C, and vice versa.

The head joint changes its relations, in which the ball represents the anterior, the socket the posterior line by a concentrated working, the other joints do so by an eccentric movement which gives an epicycloidal shape to their curves of transformation.

At the cross-lines the same movement has place, the direction in which the old (for example) right posterior line turns its posterior cross-line end is the same in which the alternate right anterior line will turn its anterior end, and they will both draw their ends over to the same (here the right) side, where the joint will slip, the remaining alternate line claim its traction, and the workings again become diagonal.

It is the rectifying of the cross-line, and that, finally, between the lung-lobes, which is, as it may be expressed, the sticking point in "setting-up."

§ 160. In man as has been mentioned, there is one more gathering than in the horse, viz.: that which brings the bearings of all the tractions above the collar bones, and unites the results of all the S S in the hands.

With this addition, which is only an extension of those turns at the upper corner of the sternum that unite the rear C C of the central and third S S together in the neck, the principles of motion in man are precisely similar to those in quadrupeds, and we may refer to the foregoing discussions for every explanation that may be required.

The great cause of deformity in civilized man we assume to be the preponderating exercise of one set of diagonal corae counteractions, until the muscles affected by them have acquired an undue proportion of strength, and permanently fixed the convexes which accompany their gatherings. Thus, the change from one set of diagonal appuis to the other no longer carries with it that complete change of socket bearings, throughout the joints of the body, which should take place. In fact, the movements of a man under such circumstances, whether walking or running, are, to a greater or less extent, varying with individuals, under the conditions of those of a horse in the canter, and in most men these conditions are those of a canter to the right, viz., on the left leg as a principal appui, and with the right arm as the " leading limb."

We now propose, first, to describe the course of action in setting-up on the *basis of the halting of a horse from the trot.*

We shall then allude to the movement from other conditions, and shall also give some *exercises founded on the filling and exhausting of the lung lobes*, and the movements which the extending influence of these brings on in the composite and the bicomposite cycles.

The lung exercises will afford the best clew for those who, from want of anatomical knowledge, or from the want of clearness in our own explanations, may find difficulties with the other methods. In addition to these advantages it is, perhaps, safest always first to fill up the lungs in any exercises which involve lifting, or give a strain to one particular part of the body.

§ 161. The movements of setting-up are not so complicated as they might seem, since the continuance of the initiatory motion entails all the others, and the chief difficulty is rather to know what directions of movement are to be permitted, than what ones are to be made.

§ 162. It may, however, be as well here to *recapitulate the leading points* assumed for locomotion and for halting.

First.—The winding-lines act for each S in two sections. The beginning and ending of each line are collateral, so that the "point of application" of an anterior "winding-line" is identical with the ending of its collateral posterior "winding-line," and the point of application of a posterior line is identical with the ending of its *collateral* anterior line. The anterior lines develope, at first, specially in their secondary or posterior sections, and are always connected with the head condyle movements, whether these be *superficial* to accommodate the turnings which ascend the neck from the ophidian and composite spines, or *thorough* to accompany the formation of the bicomposite spine.

The secondary sections of the anterior lines affect the anterior part of the eye muscles and digastrics with an outward turn; the primary, the posterior part in a similar way.

It is in displacing or equalizing the positions of the anterior cross-line ends that the primary section motion more particularly manifests itself.

The anterior lines give outward bearings on the convexes, the posterior lines give inward bearings on the *convexes*, but on the concaves they give outward bearings. Thus, when the alternate left upper lung lobe is filled in equalizing, it is with an outer bearing of the lower (left) true ribs, being in the secondary section equalization of the left post point of application.

Second.—The posterior lines would develope in a manner similar to that of the anterior lines, viz.: first, and more especially, in the front C C C, as the anterior do in the rear C C C, and end by moving the posterior cross-line ends, were it not that, in quadrupeds and man (and, we suppose, in most species of the snake), the *normal* position of the ribs for an outer bearing so interferes with their course as to bring the first development of the posterior lines also into the rear C C C, this being, however, a reflex action from the suppressed movements of their secondary sections in the front C C C. Thus it is not until the working in the front C C C is carried out that the posterior point of application can act directly in its primary section and move the posterior cross-line end.

This peculiarity in the action of the posterior winding lines is what secures progressive or (changing the appuis) retrogressive locomotion instead of two springs, one to the rear and the other forward and centering in the cross-line (§§ 31, 32).

Third.—As every stage in the process of halting is formed by induction from the continuance of the *first* action, it follows that this first action, whether of an anterior "point of application" by the head-condyle movement, or of a posterior "point" by its socket movement, must also be the *last* action, so far as induction can carry the movement. Thus, in forcing the left head condyle and its anterior line (§ 71) the equalizing of the condyles must be its final action; and in forcing the right socket movement with the right posterior action the equalization of the sockets must be its final action. But, in the first case, though the left anterior line would be reduced as the right condyle came into place, by the drawing of the right sternomastoid at the right upper corner of the sternum, a further and *separate* action of the right posterior line would be required to give full equalization to the alternate left posterior line and carry out the collateral drawing from the pelvis. And in the same way, in the second case, induction from the right posterior line being finished by equalizing the position of the left socket with the right one and drawing on the left shoulder-blade, a *separate* action of the left anterior line would be necessary to carry out the drawing of the right alternate line from the head.

Fourth.—Although the forcing tractions may be begun at any point, and their relations afterwards adjusted, yet the regular succession, in order to a smooth working, is to begin in any S with the point of application, either of the anterior or posterior line, and to continue that line through the formation of all the "spines" before the other line begins by induction.

Fifth.—In the equalizations, the new formations

are evidently developments on the *alternate* lines, and we may therefore, instead of *forcing* the old lines in order to form the equalized alternates, begin by forming the alternates and *drawing* from the old lines. That is, instead of forcing A and B in order to produce A and B, begin with forming A and B and reducing A and B.

In this case of beginning with the alternates, however, the direction of action will be, at every point, reversed. For instance, in producing b'^1 from b'^1 the right true ribs are drawn around by the right to the rear, being the movement in the secondary section of the right posterior lines; but beginning with b'^1 these true ribs (the "right shoulder") are advanced, because the basis of action is the *left* posterior line, which draws back the left true ribs.

Sixth.—The general course of the winding lines of the *bicomposite* spine may be given as follows, and since the object in Setting-up is to form these lines by the fusing of those of the S S S, their course should be thoroughly apprehended:

The *left anterior winding line in the bicomposite spine* passes from the left side of the head, through the sterno-mastoid to the left upper corner of the sternum, thence, around the right side of the body to the left hip joint in rear.

It will be seen that the sternum here contains the anterior line elements of the ophidian and third S S. With these the lines of the neck S are fused through the sterno-mastoid muscles and the front muscles connecting the pelvis and the posterior end of the sternum. These having respectively received the counteractions of the two C C of the neck and third S, straighten in double twist and fuse them with the ophidian S in the bicomposite spine.

The right anterior winding line passes from the right side of the head, through the sterno mastoid muscle to the right upper corner of the sternum, thence, around the left side of the body to the right hip joint in rear.*

The right posterior winding line passes from the right hip-joint, in front, through the right front abdominal muscles to the right side of the lower end of the sternum, thence, around the body by the left side to the right shoulder blade, thence, by the right neck shoulder blade muscles to the back of the head.

---

*In man the connection by the collar bone gives an action for each anterior line on the corresponding shoulder. In the horse the levator humeri furnishes this connection.

The left posterior winding line passes from the left hip joint, in front, through the left front abdominal muscles to the left side of the lower end of the sternum, thence, around the body by the right side to the left shoulder blade, thence, by the left neck shoulder blade muscles to the back of the head.

Seventh—.The centre between the upper and lower lung lobes is the focus of force, and all "setting up" is directed to centering the final gathering on this point, and therefore "*all straightening of the figure is concentrated between the shoulder-blades,* and not at the small of the back.

The *seat of the "cross lines"* and of each *ground* of appui, artificial or real, then would be—

For the *ophidian spine* between the upper and lower lung-lobes; having for its artificial ground the sternum and the eye balls.

For the *composite spine,* at the small of the back, behind the diaphraghm; having for artificial ground the upper end of the sternum, when the anterior winding line forms, and the lower end when the posterior; and, in a measure, taking in the front of the pelvis and the shoulder sockets.

For the *neck S,* at the junction of its upper and lower C C, in man about one third down from the head joint; having for artificial ground the lower jaw, and, in a measure the shoulder sockets.

For the *bicomposite spine*, at the head joint; having for a ground the terminations of the posterior and anterior limbs, and for an artificial ground the eyeballs.

The composite spine, in forming with the anterior lines, acts first with the lower C of the neck, and therefore on the upper end of the sternum as an artificial ground; in forming with the posterior lines it acts first with the upper C of the third S, and therefore on the lower end of the sternum in that relation.

All combine, more or less, on the eyeballs, and finally the bicomposite spine joins its action with that of the ophidian and these points.

Eighth.—As has been before noticed, the *pressure* of each point of application *principally* produces the concave under it (§21 and following). Thus the anterior (left) concave is mainly due to the pressure of the left anterior point of application, and the effect of its line if carried throughout the S would be to make it wholly concave to the left.

In the same way, the right posterior pressure extended would form the whole line concave to the right.

We say " principally produces," for the active working of the (left) anterior winding line against the stationary position of the posterior end of the S, has begun the lower part of the rear C, before the active working of the posterior point begins, and the virtual effect of the (right) posterior line is similar at the front part of the front C. The concavities, however, were much increased by the pressure of the diagonal points of application.

It is the extension of the (left) anterior concave and its acceptance by the *(left)* posterior point of pressure, and the extension of the *(right)* posterior concave and its acceptance by the (*right*) anterior point of pressure, which brings about the change of curvatures, or for halting and setting up, the partial change and consequent equalization.

Ninth.—The filling of either upper or lower lung lobe brings its C forward, when, of course, the other C of the S passes relatively backwards. Thus, when the upper lung lobes fill, the lower lung lobes, and with them the hinder limbs, pass to the rear; when the lower lung lobes fill, the hinder limbs come forward.

Tenth.—The inner bearing of the propellers developed on the convexes includes the reactions of an anterior winding line and its diagonal posterior, and when equalized produces the two on the opposite side.

Eleventh.—Any point giving *off* a bearing moves in the opposite direction *by reason of the loss* of it, and vice versa. Thus, when the right shoulder blade gives off inner bearing to the left, the former recovers itself on the outer bearing. This will be particularly noticeable in setting-up by double action (§ 179).

§ 163. Since the previously formed anterior winding line is the *normal* obstacle to the *direct* action of the posterior point of application, this anterior line must be reduced before the full equalization of the posterior lines can have place.

In beginning with the (left) anterior line, the equalization of the outer bearings, the filling of the (right) lower lung lobe, and the reducing of the upper convex proceed as far as the equalization of the anterior cross line ends will carry them, but, in order to give full equalization to these lines, there remains the equalizing of the inner bearings.

This equalizing of the inner bearings has place first (by reflex action from the secondary section of the posterior line) in the posterior convex; then, at the seat of the secondary section, in the anterior convex, by direct action; and these being carried through, the posterior point of application is so brought into connection with its cross line end that it can bring into action the *alternate* (left) posterior point of application, and cause this latter to produce the filling of the (left) upper lung lobe under the left true ribs. Finally, movement in double action gives equal traction at both upper corners of the sternum.

Thus, in beginning with an anterior line, the filling of the corresponding *posterior* lung lobe leads the movement up to a^3 a'^3, and the filling of the *anterior* lung lobe takes place only just before the action in b'^3 b^3.

On the other hand, in beginning with the (right) posterior line, the equalization of the inner bearings on the convexes begins the movement; this is succeeded by the equalizing of the upper lung lobes, and of the posterior points of application, and this by the neck action, forming the bicomposite spine for the posterior lines; finally, the remaining half of the outer bearing for the anterior line is adjusted, the (right) lower lung lobe filling, and the neck action for the anterior line forming the bicomposite spine, prepares all for a movement in double action which gives the same traction on both shoulder blades as the terminating poitns of the posterior lines ; and, on the upper end of sternum for the anterior lines, as their points of connection with the head.

§ 164. We subjoin a tabulated view of the movements of setting up when commencing with the ophidian S. It will not, however, in practice be necessary to follow out the details. We shall number the methods proposed for setting up with a view to after reference.

No. 1—(Left) ANTERIOR LINE Leading.

Forcing A a^1.

Equalizes* lower (right) false ribs with (left) ones on outer bearings.

Fills (right) lower lung lobe.†

Anterior portion of (left) eye muscle turns outward equalizing with the analogous portion of the right eye muscle.

---

* When the word equalize is used, it denotes "so far as the movement in question will carry it." The final movement is necessary to *complete* any part.

† The lung lobes draw from each other collaterally—thus, the right lower lobe draws from the right upper; the left upper from the right lower.

Forcing A a$''^1$.

(Right) upper C (true rib) cavity reduced.

"Left-right" and "right-left" anterior cross-line ends (§ 20) equalized with the left-right as to position (both come to centre). This takes effect between the upper and lower lung lobes, reducing the right upper lung lobe from its posterior part.

Posterior portions of eye muscles equalize on outer bearings.

(Left) ANTERIOR LINE Leading.

Forcing A a^2.

Action a^1, continuing spreads to third S, equalizing sides of pelvis.

The consequent movement of the hip-joint sockets causes an equalization between the outer bearings of the left and right feet.

The (right) lower lung lobe continues to expand and the (right) diaphragm leaf to spread.

The anterior portions of the digastrics equalize on outer bearings; the (right) digastric passing, as it were, over the left.

Forcing A a'^2.

Action of a'^1 continuing, spreads to neck S.

Anterior cross-line ends, still keeping their focus between upper and lower lung lobes, equalize the upper corners of the sternum and of the sides of the spine, at the small of the back—diaphragm pillars—in outer bearings.

The posterior portions of the digastrics equalize on outer bearings.*

---

* The arms are affected by this movement, as the legs were by $a2$, but to trace their motions, as in the trot, will too much complicate the table.

(Left) ANTERIOR LINE Leading.

Forcing A a^3.

Action of a^1 continuing, and spreading through a^2, equalizes the hinder limbs on their outer bearings.

Head joint condyles equalize, in the thorough movement, as to their posterior end pressure.

Forcing A $a^{\prime 3}$.

Action of $a^{\prime 1}$ continuing and spreading through $a^{\prime 2}$.

The front C C of the lung tips (§ 112) equalize as the movement turns on them.

Anterior cross-line ends—still in focus between the upper and lower lung lobes, and further moving the upper corners of the sternum and the sides of spine at the small of the back — now equalize the lower jaw articulations on their outer bearings, the (right) articulation making a sort of lateral epicycloidal movement on the (left) one.

The head-condyles equalize in the forward movement across (§ 92) the sockets by inner edge.

All being thus drawn up to the head, the arm sockets equalize in outer bearing from right to left.*

---

* This action at the shoulder joints, it will be seen, changes the direction of the line of general pressure (§ 91), which up to this point, has been toward the right; and introduces, by changing it incipiently toward the left, the posterior line equalization which ends in the head joint *sockets*.

The effect of this on the head connection is to bring the stress on the right sterno-mastoid at right upper corner of the sternum.

(Left) ANTERIOR LINE Leading.

Forcing by Induction. } $B\ b''^1$.

By bringing back the "left-right anterior cross-line end" from its displacement, the "right-left posterior end" is brought into action, and by but a little further movement of a^1, as it fills the right lower lung lobe, the necessary forcing, in the right posterior line, is given, and the reaction of b^1 begins in the posterior C of the ophidian S. This equalizes the inner bearings of this C, after which the action in the secondary section equalizes the inner bearings of the anterior C.

The eye muscles equalize on the inner bearings—first, by reaction in the posterior portion; then in the anterior portions at the "secondary sections" of the posterior line (§ 25).

9*

Forcing by Induction. } $B\ b^1$.

The right lower point of application now draws directly, and at the focus, between the lung lobes, equalizes the posterior cross-line ends of the ophidian S.

The eye muscles equalize on the inner bearing at their posterior ends.

(Left) ANTERIOR LINE Leading.

Forcing by Induction. } $B\ b'^2$.

The action of a^1 continuing, the hip-joints are equalized on the inner bearings, and with the secondary section, the *lower* end of the sternum is equalized on its inner bearings against the outer bearings $A\ a^2$ of the upper end. By this the (right) false ribs gain in prominence, outward and forward, while the left draw in, as they fall into the traction of the left posterior winding-line.

The digastrics equalize on the inner bearing (the right one gaining on the left), the posterior portion first, then the anterior.

Forcing by Induction. } $B\ b^2$.

The action of a'^1 continuing, the upper left lung lobe fills from below.

Digastrics finish equalizing on inner bearing, at their rear ends.

(Left) ANTERIOR LINE Leading.

Forcing by Induction. } B b$'^3$.

Action of a$'$ continuing through b$'^1$, equalizes the hinder limbs on inner bearings by the reactionary movement with lower C of the neck S, and as b$'^3$ takes effect in its secondary section, with the upper C of the neck S, equalizes the shoulder joints.

Forcing by Induction. } B b^3.

Equalizes the lower jaw articulations, and thus ends with the stress on the right sterno-mastoid, drawing on right upper corner of sternum.*

---

* After which the finishing adjustment, bringing both corners of sternum into equal stress, is made in "double action."

§ 165. We next consider the

No. 2—(Right) POSTERIOR LINE Leading.

Forcing B b'^{1}.

The first action *felt* is in the secondary section of the (right) posterior line, which is that of the (right) true ribs, forcing their movement and pressing to the (left) as they turn outward and backward;* but the first *result* is reflexion, i. e., the lower ribs equalize their inner bearings, from the action in the secondary section. The true ribs then equalize, as the secondary section.

By the first movement the (left) lower false rib cavity is diminished; by the second the (left) upper true rib cavity is prepared for enlargement.

Eye muscles act as in § 164, same column.

* To prevent mistakes, we repeat what was before said, that if, instead of forcing the old line, the forming of the *alternate* posterior line lead the action, this movement is forward.

Forcing B b^1.

The forcing of the posterior line next brings about an equalization of the posterior cross-line ends, *so far as possible;* but from the displaced ("left-right") anterior end (§ 32) holding its corresponding posterior end from its movement to the (right), this equalization is performed with a decided "list" to the left, and this remains until the change to anterior line action liberates the displaced (left-right) end.

The focus of the cross-line end action is always between the upper and lower lung lobes.

Eye muscles make their inner bearing equalization at their posterior ends.

(Right) POSTERIOR LINE Leading.

Forcing B b'².

The action of b'¹ continuing the reaction in the lower C of the composite spine, and the action in its upper C, first equalize the hip-joints (by the movement of the pelvis) on their inner bearings; then the lower end of the sternum, with which goes the inner bearing of the digastrics, first of their posterior, then of their anterior portion.

Forcing B b².

Action of b¹ continuing.

Posterior "cross-line ends" still keeping their focus between the upper and lower lung lobes, and the "list" of the general line to the left being still maintained, equalize at the small of the back—diaphragm pillars—following which the (left) upper lung lobe fills with air from below, and the left line equalization begins in the neck-root joint.

The posterior portions of the digastrics equalize on inner bearings at their posterior ends.

(Right) Posterior Line Leading.

Forcing of B b'³.

The pressures continually collecting and swelling at the anterior end of the left head joint socket, equalize with the similar point of the right socket, so far as the displaced anterior cross-line end allows.

The shoulder-joints equalize, the right giving off to the left, on which latter the stress of the movement decidedly comes.

Forcing B b³.

The lower C C of the lung tips equalize as the head joint, with the upper C of the neck S, allows the turning of the outer edge of its sockets.*

The lower jaw articulations equalize, changing for this succession of movements (as at A a³ the shoulders did for the former one) the general line of bearing. But it is to be remembered that all the lines gained are to be held, and that the "induction" of the anterior line will make its own alterations.

---

* The pressure of the left socket should diminish as it equalizes with the right (both being reflected pressures); but the displaced cross-line end still holds back the movement.

(Right) POSTERIOR LINE Leading.

Forcing by Induction. } A a′.

The inner bearing belonging to the right side of the posterior C C and to the left side of the anterior C C, has already been transferred to them, they now receive their portion of the outer bearing.

The right lower lung lobe fills.

The eye muscles equalize in front on outer bearings.

Forcing by Induction. } A a″¹.

The anterior cross-line ends equalize between the lung lobes, and to a certain extent, free the posterior ones, to assume the positions from which they have been hitherto restrained.

Eye muscles equalize in their rear portions on outer bearings.

(Right) POSTERIOR LINE Leading.

Forcing by } $A\ a^2$.
Induction. }

Hip-joints equalize on outer bearings.

Digastrics equalize in front portions on outward bearings.

Forcing by } $A\ a'^2$.
Induction. }

Upper corners of sternum equalize.

Digastrics on outer bearing at posterior portions.

Adjustment of crossline ends at small of back—diaphragm pillars.

(Right) POSTERIOR LINE Leading.

Forcing by Induction. } $A\ a^3$.

  Actions of b'^1 continuing. Equalization of lower jaw articulations finished from b^8.
  Hinder limbs equalized throughout.

Forcing by Induction. } $A\ a'^3$.

  As the right head condyles equalize, the shoulder joints do the same. The right shoulder passing outer bearing to the left one, on which, as the right condyle comes into place, the left rear neck muscles draw,* completing the left posterior winding line.

---
*Finishing movement, by which rear neck muscles are equalized, is made in "double action."

See Appendix II.

§ 166. We have given the table for setting-up, when commencing with the ophidian S, in pretty full detail, but as was said, it is not necessary in practice to follow these details mentally, since one produces the other, if allowed.

Two chief points are to be kept in mind, viz., that the movement of the cross-line ends more particularly changes the curvatures, and that the focus of this movement must be maintained between the upper and lower lung lobes, that is, at the centre of the ophidian spine. Each motion being brought into this point, whether from the small of the back or from either end of the sternum, etc.

§ 167. No. 1.—For "*setting-up*" *with the (left) anterior line of the ophidian S leading, as in the first table*, it seems only necessary to observe the following points :

The chin, while not initiating the left head condyle movement, must be kept sufficiently up, so that no clamping of the head joint in the opposite movement shall interfere with the current one.

The movement being once initiated from a "point of application" at the base of the neck-root joint on the (left) side, particular attention must be given to the two fundamental motions, i. e., the filling of the right lower lung lobe, and the *subsequent* equalizing of the anterior cross-line ends between the upper and lower lung lobes.

Next comes the extension of these to the pelvis and hip joints, and the subsequent equalizing action at the upper corners of the sternum, and at the small of the back, both brought to bear between the lung lobes.

Then the anterior line movement *proper* ends with the lower jaw equalization, that of the head condyles with " the thorough movement," and that of the shoulder joint *initiating* the inducing of the (right) posterior line with the continuance of the first mentioned fundamental motions of the anterior and the change of the general line of pressure.

The *induction* of the equalization of the posterior lines comes, first, from the continued filling of the (right) lower lung lobe, this reduces the (left) lower lung lobe cavity; and next from the continued action in equalizing the anterior cross-line ends, this causes the equalization of the posterior ones, first, between the lung lobes, then at the small of the back, and at the lower end of the sternum, fills the (left) upper lung lobe which reduces the (left) lower lobe; lastly, the continued filling of the right lower lung lobe, equalizes the shoulder blades, and establishes the (left) posterior line at the left shoulder blade (§ 162 sixth); and the equalizing of the cross-line ends brings into equal action the two jaw articulations, the inner bearing passing from the left to the right, and leaving the stress of the head condyle movement at the right upper corner of the sternum.

§ 168. No. 2.—And for "*setting up*" *with the (right) posterior line of the ophidian S leading.* Since the rear "point of application" cannot, at once, exercise its direct working, as did that of the anterior winding line, the fundamental points are to begin the influence of the "secondary section" of the (right) posterior line by pressing the (right) true ribs to the left, at the same time that they rotate outward and backward,* producing the effect of equalizing the inner bearing first in the false ribs (posterior C), as a reflected action from the upper true ribs, on which follows equalization of this bearing at the true ribs, as the secondary section *proper;* subsequently to allow the direct action of the (right) posterior "point of application" between the upper and lower lung lobes, where, for the ophidian S, it takes effect more es-

---

* Forward, if forming the alternate posterior line first.

pecially in reducing the cavity of the chest under the (left) false ribs.

Next, for the composite S, with the extension of the ophidian S movement, comes as a reflected action the equalization of the inner bearing for the pelvis at the hip joints; then, as the proper secondary section, the filling of the left upper lung lobe from below; then, for the posterior "point of application" the inner bearing equalization at the small of the back, and at the lower end of the sternum, which last brings the focus back to the space between the upper and lower lung lobes.

As the movement by the right posterior winding line takes effect for the bicomposite spine, the reflected secondary section action equalizes the hind-legs on the inner bearings; the secondary section proper the shoulder blades; and the drawing of the (right) posterior point of application the lower jaw articulations; during which the head joint sockets equalize so far as they can, for the general line still inclines to the (left). The whole turns on the upper C C of the lung tips. The equalization of the lower jaw articulations changes the movement for the induction of *action by the anterior winding line.*

The induced equalization of the lower ribs on their outer bearings (the secondary section of the anterior line), fills the right lower lung lobe; then the action of the primary section, as the anterior cross line ends equalize between the lung lobes, begins to release the (right-left) posterior cross-line end and the general line of the body gains toward the (right).

As the action of the secondary section extends to the composite spine (the right lung continuing its filling), the right hip joint obtains a more equal bearing; as that of the primary extends, the left upper lung lobe fills more completely, and is nearly fully expanded as the equalization of the anterior cross line ends at the small of the back and the upper end of the sternum, releases the posterior line ends at the small of the back and the lower end of the sternum, all again centering between the lung lobes.

The bicomposite spine movement equalizes for its secondary section the bearings of the hinder limbs, for the cross lines the lower jaw articulations, and for its final turn the shoulder blades; leaving a final stress from the back of the head on the left shoulder-blade, which is to be adjusted with the other by a slight movement in double action.

§ 169. No. 3—The "forcing of the winding lines"

may, perhaps, be more advantageously carried out on the *pacing movement* than on that of the trot just described. If we *begin with* the *anterior lines of the bicomposite spine*, i. e., the pace in progression, the successive motions would be as follows:

First.—The movement of the (left) head condyle equalizing with the right at its posterior end, brings on the equalization of the lower jaw articulations, the (right) articulation being brought (§ 166) with a sort of lateral epicycloidal movement over the (left); thence the movement extends to the equalization of the hind-limbs on the outer bearing; thence, passing more inward to the equalization of the secondary section of the anterior line in the composite spine, viz., the pelvis, the false ribs, the filling of the (right) lower lung lobe; the secondary section in the ophidian spine bringing the action between the upper and lower lung lobes.

Second.—The equalization of the shoulder-blades on the outer bearings as the head condyles equalize on their cross joint movement, brings on the equalization of the other primary sections in the successive spines. The sternum corners and the small of the back are effected by the anterior cross-lines of the composite spine, and, finally, the equalization of these end between the upper and lower lung lobes.

Third.—As the head joint sockets begin their equalization at the forward ends, the shoulder-joints would equalize as representing the secondary section of the posterior line action, in the bicomposite spine; but the "reflected action" which takes place in the hinder-limbs must first be carried out by their equalization on the inner bearing; then succeeds the similar equalization of the shoulder-joints; next, for the composite spine, comes that of the false ribs, then of the true ribs, and the filling of the (left) upper lung lobe, from the left lower; finally, the equalizing of this section in the ophidian S.

Fourth.—The head joint sockets equalize on their outer edges; the lower jaw articulations follow, bringing on the equalization of the posterior cross-line ends in succession; the lower end of the sternum and the small of the back for the composite S; the working between the upper and lower lung lobes for the ophidian.

§ 170. It will be observed that *the lower jaw articulations go with the posterior C C in each case; the shoulder-blades with the anterior C C; that is, the lower jaw goes with the secondary section of the anterior lines, and, with the primary section of the posterior, while the shoulder-blades go with the secondary sections of the posterior, and the primary of the anterior.*

It will be also observed that the S S were carried out by halves.

§ 171. No. 4.—If for *setting-up* we begin with the *posterior lines* of the bicomposite spine in the *retrogressive pacing action*, the successive motions would be as follows: *First.*—For the bicomposite spine, the hinder limbs in reflected action, then the shoulder-blades as the *proper* secondary section of the posterior line; the head-joint sockets equalize at their anterior ends, next, for the composite spine, the false ribs; then the true ribs and filling of the left upper lung lobe; lastly, equalizing of this secondary section in the ophidian S. *Second.*—Head-joint sockets equalize on their outer edges, the lower jaw articulations equalize on the inner bearing, succeeded in the several spines by the equalizing of the posterior cross-line ends as the (right) posterior point of application draws. *Third.*—The head condyle equalization at the rear ends, and for the bicomposite spine, the lower jaw articulations equalize on the outer bearing, then the hinder limbs; for the composite spine, the pelvis, false ribs and filling of the (right) lower lung lobe, action between upper and lower lung lobes. *Fourth.*—Head condyles equalize on their cross joint movement, equalization of shoulder-blades for bicomposite spine; *next*, the upper end of sternum, small of back; and between lung lobes for composite and ophidian spines.

§ 172. We have gone through the details of equalizing the tractions of the winding lines by forcing beyond their limits those already formed, this seeming the best way of explaining the subject; but often, perhaps generally, the most practical method for "setting-up" is forming the alternates and discharging in their "wake" the old lines of the composite and ophidian spines, at the same time limiting the action to these spines, which will allow of equalizing each line after its formation.* This being done, next forming successively, in the same way, the alternate lines of the bicomposite spine, discharging the old ones and equalizing the new ones.

No. 5. To equalize the winding lines, *the formation of the (right) alternate anterior line leading.* As has been previously remarked, the bearing of the general line and the rear and front direction of the leading moving points will be reversed from those of the already given examples, because these are now those of the new line forming and drawing,

---

* Did the movement extend through the neck-root joint to the bicomposite spine we should have the full alternates.

instead of those of the old line forcing its actions and, so to speak, pushing. (§ 162. Fifth.)

As in all cases, the movement may be begun at various points. We shall commence with the (left) eye.

First.—With the (left) eye muscle turning *outward* in its *anterior* portion, commence the formation of the alternate (right) anterior line in its secondary sections, viz: The filling of the (right) lung lobe, followed by the reduction of the left false-rib concavity and the increased outer bearing at the (right) hip-socket, followed by a reduction of the (left) hip-socket. Then, with the outward turn of the (left) eye muscle, in its *posterior* portion, commence the formation of the primary sections, viz.: the gaining between the (right) lower and left upper lung lobes of the (right-left) cross-line end to the (left), followed by reduction of the position of the (left-right) end to the (right); gaining of the same point at the small of the back to the right followed by the reduction of the (left-right) point to the right,* a similar gain and reduction between

---

* In many cases, from the great displacement of the left-right anterior cross-line end to the left, its restoration is the main feature. The right side at the small of the back straightening its incurvation toward the left in a very marked degree. It may be remarked here, that the new points rise above the old ones, and also pass (in man) in front of them, corresponding to the passage of the concaves under the convexes in the snake (§ 60).

the (right and left) upper corners of the sternum.* Continuing the drawing will equalize the whole.

Second.—With the (left) eye muscle turning *inward*, first, in its rear portion, by reflected action; lastly, in its anterior portion, by proper action, allow the alternate (left) posterior winding-line to form for the secondary sections, viz., by bringing the inner bearing on the (right) false ribs, which implies a reduction of the cavity under the (left) false ribs (§ 39 and note), by taking the inner bearing from the (left) hip joint to the (right); by filling the (left) upper lung lobe from below as the lower (left) true ribs turn outward. Then for the primary section, the drawing of the alternate anterior (right-left) cross-line end will bring the alternate posterior (left-right) end into action; discharging the old point between the lung lobes; at the small of the back; and at the lower end of the sternum.

Third.—From the bicomposite alternate (right) anterior line, first, on the right lower jaw articulation, then on the left shoulder blade, discharging the old line—equalize. Then the bicomposite alternate (left) posterior line, on the left shoulder

---

* It is often necessary to equalize somewhat the small of the back and corners of the sternum before the movement *between* the lung lobes can be effected.

blade, then on the right lower jaw articulation, discharging the old line—equalize.

§ 173. *Remarks.*—As the alternate line is the "new line" of the previous methods, the "anterior line leading" movements end here, as there, with a stress on the same point, viz., the right upper corner of the sternum.

In all these movements, where the anterior line leads, the order of succession for the cross-line ends in the composite spine is small of the back, upper corners of sternum for the anterior ends; small of back, lower end of sternum, for the posterior ends. This is a necessary consequence of the throughout continued filling of the (right) lower lung lobe, as the conspicuous foundation of the movement. On the other hand, for the leading of the posterior line where the inward pressure of the (right) upper true rib convex, whether forward or to the rear, is the foundation of the movement, the succession is necessarily lower end of sternum, small of back for the posterior ends; upper end of sternum, small of back for the anterior ends.

§ 174. No. 6.—To equalize the winding lines, *the formation of the alternate (left) posterior line leading.* As we cannot begin with a rear point of application (§167), and, therefore, not with its representative, the rear point of the (left) eye muscle; and, as we cannot develope the secondary section of the alternate posterior line without some corresponding point d'appui, in the old line, we take the right upper ribs as this point d'appui, and begin by forming against it, the secondary section of the alternate (left) posterior winding line; first in its "reflected" action of drawing the inner bearing from the (left) false to the (right) false rib articulation; and, then, in its "proper" action from the (right) true to the (left) true ribs.

It will be remembered (§171), that the movement of every point is, now, the reverse of what it was in the forcing of the old line.

We shall make the description of this movement in more *general* terms than have been employed heretofore.

In the composite spine. First.—Pressing the upper part of the (right) upper ribs forward and to the (left), an appui is taken on that action in the course of the alternate (left) posterior line, which will, for the *secondary section* (by reaction), draw the inner bearing from the (left) false ribs to the right ones, then (by proper action), that of the right true ribs to the left ones, and begin to fill the left upper lung lobe in its lower portion as the left true ribs take the inner bearing.*

Next, for the *primary section*. Developing the position of the alternate (left-right) posterior cross-line end discharge the old (right-left) posterior—straightening the right side of the spine at the small of the back, and releasing the traction on the left side of the lower end of the sternum.†

Second.—The continuation of the movement brings on a retraction of the (right) upper point of application for the ophidian spine; the filling of the right lower lung lobe for the *secondary section* of the anterior line. Then, for the *primary section*, the formation of the new, and discharge of the old-line, between the right and left upper corners of the sternum, and the same for the two sides of the small of the back.

In the bicomposite spine. First.—The left shoulder-blade taking from the right shoulder-blade, then, the right jaw articulation from the left one for the alternate posterior line.

Second.—And, for the whole, *lastly*. The right jaw articulation taking from the left one; then, the left shoulder-blade taking from the right shoulder-

---

* This lung lobe does not fill until the preceding movements have been well carried out.

† This traction § 39 belongs to the old (right) posterior winding line.

blade for the alternate anterior line; the whole ending with a stress on the (left) shoulder-blade.

These last bicomposite actions move the neck-root joint and bring the focus of force to the space between the upper and lower lung lobes.

§ 175. Nos. 9 and 10.—§ 177. The forming of the *alternate lines* in the pace developement, followed by their equalization after the discharge of the old lines.

No. 9—For the (right) *anterior* alternate line leading, we have, *first*, the attempt at rotation backward and inward of the (right) head condyle; then the incipient convex on the (right) of the lower C of the neck S, followed by the discharge of the (*left*) upper corner of the sternum; then the formation on the outer bearing, for the (right) jaw articulation, followed by the discharge of the left articulation, then the developement of the (right-left) anterior cross-line end in its position, and the reduction of the (left-right) followed by the formation for the left shoulder-blade on the outer bearing, and reduction of the right.

For the (left) *posterior* alternate line *by induction;* *first*, inner bearing formed on the right of the lower C of neck S (reaction); reduction of left side; inner bearing on left of upper C (proper action);

reduction of right side; formation of left shoulder-blade on inner bearing; reduction of right shoulder-blade; then the development of (left-right) posterior cross-line end, and reduction of the (right-left); followed by the formation for the (right) jaw articulation on the inner bearing, and reduction of the (left). Finally, drawing of (right) sterno-mastoid.

§ 176. No. 10.—For the (left) alternate *posterior* line *leading*. The most marked points seem to be the forming pressure of (right) socket gaining its pressure at the front part of its condyle—and the reduction of the pressure of the (left) socket, as if it were separating from its condyle.

For this (left) *posterior* alternate line leading by forming pressure of the (right) socket, we have, *first*, the inner bearing on the (right) side of the lower C of the neck S (reaction); reduction of left side; inner bearing on left of upper C (proper action) &c. &c., being the repetition of No. 9, excepting that the posterior line comes first and depends not on induction, but on the forming socket-pressure; and that the anterior line comes second by induction, working from the right lower jaw articulation as the posterior line continues its action.

10*

Nos. 7, 8, 9 and 10 run very much into each other. In fact, in 9 and 10 the equalization of the composite spine follows on the first part of the second set of lines in action whichever this may be. Thus, in No. 10, so soon as the left shoulder-blade moves in connection with the lower C of the neck, the right shoulder-blade steadies the upper C of the composite spine for its reactionary action in the lower ribs.

PRACTICAL SETTING-UP.

§ 177. Of the EIGHT METHODS OF "SETTING-UP," which have just been enumerated, THE BASES may be described as follows, and with the descriptions already given we shall refer more particularly to the present section for *practical explanation*. The process of exhausting and filling the lung lobes will be discussed more fully under "setting-up with double action;" but it may be remarked here that the mouth should be kept closed, the air passing out through the nostril collateral with the exhausting lobe, and entering by the nostril collateral with the filling lobe. It will not, however, be necessary to attend to the latter process, as a good part of the filling will come from transfer from an upper to a collateral lower lobe, or vice versa. The paragraph concluding No. 1 applies with obvious modifications to all the numbers.

We have taken no notice of the movement of the head condyles or eye muscles, as they will adjust of themselves.

It is to be remembered that the upper lobes fill or exhaust from below, the lower lobes from above.

No. 1.—Forcing the anterior line in the trotting movement. (1.) Exhausting the lower (left) lung lobe until this action brings on the filling of the lower (right) lobe. (2.) By induction from (1.) and after allowing the necessary connections of the neck and pelvis to adjust themselves. Exhausting the upper (right) lung lobe, until this brings on the filling of the upper (left) lobe, and fresh adjustment of the pelvis and neck.

The movement would be begun by drawing back the left upper part of the chest near the first rib. As to the adjustments, these will be—(1.) the (left) lower jaw articulation giving off its inner bearing as the lower (left) lung exhausts, and the (right) jaw articulation taking outer bearing as the lower (right) lobe fills. (2.) The (right) shoulder joint giving off inner bearing as the upper (right) lung lobe exhausts, and the (left) shoulder joint taking outer bearing as the upper (left) lung lobe fills. Consequent on the exhaustion of the lower C of the right lung tip* and filling of the lower C of the left, being the final action, the right sterno-mastoid then draws from above on the right upper corner of the sternum.

It will be observed that in these movements the lower (left) false rib articulations cross the central line of the body; both they and the (left) hip joint passing in front, across the body to the (right) and falling outward, as they give off inner bearing. The (right) true ribs and right shoulder blade pass *relatively* to the (left), beginning below, but the uppermost true rib is not adjusted until the body S being virtually carried through its movement, the right shoulder joint moves with the neck S, completely separating the sterno-mastoids in front.

No. 2.—Forcing the posterior line in the trotting movement. (1.) Exhausting the upper (right) lung lobe, until this action brings on the filling of the upper (left) lobe. (2.) By induction from (1), and after allowing the pelvis and neck connections to adjust themselves. Exhausting the lower (left

---

* The upper C of each lung tip goes with its diagonal lower lung lobe; the lower C with the upper lobe, to which it is attached.

lung lobe, until this brings on the filling of the lower (right) lobe and fresh adjustments of the neck and pelvis.

Since the posterior point of application cannot, at first, act directly, the movement would be begun by drawing backward the (right) upper part of the chest, near the first rib, but at the same time allowing this to press inward, to accommodate the reflected action in the (left) false ribs. As to the adjustments, these will be (1) the (right) shoulder joint giving off its inner bearing as the upper (right) lobe exhausts, and the left shoulder joint, taking outer bearing as the upper (left) lobe fills. (2.) The (left) jaw articulation giving off inner bearing, as the left lower lung lobe exhausts, and the (right) jaw articulation taking outer bearing as the lower (right) lobe fills. Consequent on the exhaustion of the upper C of the (right) lung tip and filling of the upper C of the (left) lung tip being the final action, the left head muscles then draw from above on the left shoulder blade.

No. 3.—Forcing the anterior line in the pacing movement. (1.) Begins with the (left) jaw articulation giving off inner bearing with exhaustion of lower (left) lung lobe. The movement of the neck S corresponding to this, brings around the right shoulder joint, giving off inner bearing with exhaustion of the upper (right) lung lobe. (2.) The neck motion continuing, the (right) jaw articulation takes outer bearing with filling of lower (right) lung lobe, and the neck movement still continuing, the left shoulder joint takes outer bearing with filling of upper left lung lobe. Finally, although differently from the trot, the drawing of the right sterno-mastoid is in advance of the equalization of the lung tips, yet, consequent on the exhaustion of the lower C of the (right) lung tip, and filling of the lower C of the left, being the final movement, the drawing of the upper (right) corner of the sternum is from above.

No. 4.—Forcing the posterior line in the pacing movement. (1). Begins with the (right) shoulder joint giving off inner bearing with exhaustion of the upper (right) lung lobe. The movement of the neck S corresponding to this, brings around the (left) jaw articulation, giving off inner bearing with the exhaustion of the lower (left) lung lobe. (2). The neck motion continuing, the (left) shoulder joint takes outer bearing with the filling of the upper (left) lung lobe, and the neck movement still continuing, the (right) jaw articulation takes outer bearing with the filling of the lower (right) lung lobe. Finally, the drawing of the (left) side back muscles of the neck, moving in advance of the equalization of the lung tips, consequent on this and on the exhaustion of the upper C of the (right) lung tip, and filling of upper C of the (left) lung tip, being the final action, the drawing on the (left) shoulder-blade is from above.

No. 5.—Forming first the *alternate* anterior line in the trotting movement. (1). Filling the lower (right) lung lobe. Exhausting the lower (left) lobe. (2). Filling the upper (left) lobe. Exhausting the upper (right) lobe.

The jaw articulations and shoulder joints, as before, follow the lung movement, the (right) jaw articulation taking outer bearing, then the (left) one giving off inner bearing. The (left) shoulder taking outer bearing, the right shoulder giving off inner bearing. Then, as the lung tips equalize, the jaw, and lastly the shoulders, take a final adjustment, ending with a drawing at the right upper corner of the sternum, which, instead of coming as in No. 1, from above, by the sterno-mastoid, now comes from below, along the right side of the sternum.

No. 6.—Forming first the *alternate* posterior line in the trotting movement. (1). Filling the upper (left) lung lobe. Exhausting the upper (right) lobe. (2). Filling the lower (right) lung lobe. Exhausting the lower (left) lobe.

Referring to the first part of the concluding paragraph of No. 2. (1). The (left) shoulder joint takes an outer bearing as the upper (left) lobe fills, and the (right) shoulder joint gives off inner bearing as the upper (right) lobe exhausts. (2). The (right) jaw articulation takes an outer bearing as the lower (right) lobe fills, and the left jaw articulation gives off inner bearing as the lower (left) lobe exhausts. As the lung tips equalize, the jaw, and lastly the shoulders, take a final adjustment, ending with a drawing on the left shoulder-blade from below, instead of, as in No. 2, from above.

No. 7.—Forming first the *alternate* anterior line in the pacing movement. (1). Lower (right) and, succeeding this, the upper (left) lung lobe fills. (2). Lower (left) and, succeeding this, the upper (right) lung lobe exhausts.

The head being raised, the (right) lower jaw articulation commences the movement, taking on outer bearing, as— following the movement of the jaw—the (right) lower lung lobe fills. Next, the left shoulder-blade, which meantime must not be kept from following the general movement of the neck, takes on outer bearing as the (left) upper lung lobe fills. The (left) jaw articulation gives off inner bearing as the (left) lower lung lobe exhausts, and the left hip joint is released. The (right) sterno-mastoid now draws at the (right) upper corner of the sternum. Finally, the (right) shoulder joint gives off inner bearing as the right true ribs adjust themselves, and the force along the (right) side of the sternum coming from below, unites with that of the right sterno-mastoid in making the tractions collateral.

No. 8.—Forming first the *alternate* posterior line in the pacing movement. (1). Upper (left) and, succeeding this, the lower (right) lung lobe fills. (2). Upper (right) and, succeeding this, the lower (left) lung lobe exhausts.

The (left) shoulder joint takes on outer bearing as, following the movement of the shoulder joint, the (left) upper lung lobe fills, the (right) jaw articulation, meantime, following the general neck movement. The (right) jaw articulation takes on outer bearing, and the (right) lower lung lobe fills. The (right) shoulder joint giving off inner bearing, the right true ribs are released, then, the (left) jaw articulations giving off inner bearing, the left false ribs and the left hip joint. Lastly, the drawing on the left shoulder-blade from the head having been already established, the line of traction is made collateral by a drawing from below.

§ 178. If we have previously succeeded in explaining the successive movements of the parts in detail, the fillings and exhaustions given in this section—the action of the leading one being continued throughout—should indicate those movements. The key to "setting up," as before mentioned, is rather the knowing where to yield than where to initiate movement.

The pacing movement ought, perhaps, theoretically, to equalize a C of the lung tips before its corresponding lower or upper lobe; but in practice, particularly where there is even the slightest distortion, we think it will be necessary to give some precedence to the lobes.

§ 179. There remain two points to be spoken of. These depend upon the fact that when the shoulder-blades are turned outwards they act upon the lower part of the neck-root joint, and affect the trunk; but when they are turned inwards they act upon the upper part of this joint and affect the neck S.

By pressing the right shoulder forward, and to

the left, across the neck, we bring on motion in the secondary section of the left anterior winding line, and thus initiate a forcing of the action of the (left) head condyle.

By pressing the left shoulder forward, to the right, across the neck, we bring on motion in the line of reaction of the secondary section of the right posterior winding line, and thus initiate a forcing of the indirect pressure of the left head socket (§39).

§180. *Equalization on two lines at once* (or double action in halting). In this, a similar interchange of action and of bearings, from side to side, takes place as when the lines act singly, but the "shapes" do not show themselves laterally. It finishes also in the same result of bringing the tractions into collateral, instead of diagonal connection. As before, the eye muscles accompany more especially the ophidian, the digastrics the composite the lower, jaw the bicomposite spine.

§181. No. 11.—*Double action equalization, the anterir lines leading.*

On the *anterior lines* (leading).—Ophidian spine—by outward bearing action, in lower C, of central S, both lower lung lobes begin to fill as *scecondary section action.* Then both anterior cross-line ends exchange for collateral actions, between the upper and lower lung lobes, as *primary* section action. The changes caused in the lines of gravity give this fundamental movement a general effect throughout the body.

Composite Spine.—The lower lung lobes go on filling and the hip-joints interchange outer-bearings for the *secondary* section. Then the small of the back and the upper corners of the sternum adjust their right and left sides, as continuing the action of the anterior cross-line ends for the *primary* section.

Bicomposite Spine.—The lower lung lobes continuing to fill, the hinder limbs adjust themselves for the *secondary* section, the lines passing in rear of the hip joints. Then, the lower jaw articulations adjusting on the outer bearings, followed by the rotary movement of the head condyles, on their rear ends, and the forward across-joint movement of the condyles, followed by the adjustment of the shoulder-blades on the outer bearings, represent the anterior cross-line ends in the *primary* section.*

---

* The movement between the lower jaw articulations and shoulder-blades, crossing at the junction of the neck C C, is as if the body swung between them, as the bicomposite tractions confirm those which have

On *posterior lines* (by induction). Since the last point in the development of the anterior lines was a bearing on the shoulder-blades, some action on these will result from each stage in the subsequent development of the posterior lines until they bear on the lower jaw.

Ophidian Spine.—Interchange of inner bearings between right and left sides of lower ribs begin to straighten these ribs in double twist, and thus complete the filling of the lower lung lobes (reaction). Interchange of inner bearings and between sides of the upper C, (proper action,) and commencing filling of the upper lung lobes from their lower portions for *secondary* action. Then, posterior cross-line ends adjust themselves between the lung lobes for *primary* section.

Composite Spine.—As the interchange of inner bearings between the lower C ribs continues from the reactionary movement, its effect extends in the muscles from the lower end of the spine to the front of the pelvis, and does not yet reach the hip socket. Extended interchange between lower ribs of inner bearings, which adds to filling of lower

---

gone before. This movement changes the "general lines of thrust" against the edges of the head joint sockets.

The lung tips (second C) movement occurs with the two condyle movements.

lungs and causes an adjustment between the muscles passing from lower end of the sternum to the pelvis (reaction). Interchange of inner bearings between rib articulations of upper C and filling of upper lung lobes, for *secondary* section. Then, adjustment of posterior cross-line ends at small of back and lower end of sternum, as continuing action of posterior cross-line ends* and drawing on posterior points of application (which we may perhaps consider to be placed at the front projecting points of the hip bone) for the primary section.

Bicomposite Spine.—Hinder limbs adjust themselves, as to inner bearing, on front part of feet (reaction). The shoulder-blades adjust themselves, followed by the movement at the anterior ends of the head sockets (proper action). Then the movement backward of the sockets along the outer edges of the condyles, followed by the adjustment of the jaw articulations, as representing the—on this bearing—posterior cross-line end adjustment; and, finally, the adjustment and drawing of the posterior points of application at the front of the hip sockets and down the hinder limbs, for the *primary* section.

---

* We have not thought it worth while to speak of the movement of the diaphragm and its pillars.

Stress on upper corners of sternum. Upper C of lung tips pivot for head socket movement.

§ 182, No. 12.—*Double action equalization, the posterior lines leading.*

On the *posterior lines* (leading) *ophidian spine*. Interchange between sides of lower C of central S of inner bearings (reaction). This action, which, in the preceding formation (§ 181), when superinduced on the outer bearing, completed the filling of the lower lung lobes, by straightening the ribs in "double twist;" now, on the contrary, having only the effect of pressing the ribs together, drives the air from these lobes. Interchange of inner bearings between the sides of the upper C of the central S (proper action). This action of the posterior points of application draws the rib heads down, and thus raising the outer part of their arcs enlarges the upper lung lobes. This for the secondary section. For the primary section the posterior cross-line ends separate between the lung lobes, and, following this, the posterior points of application, do the same, at the lowest dorsal vertebræ.*

Composite Spine.—Again, the lower ribs are com-

---

* What are "points of application" for the "ophidian spine" come to be cross-line ends for the "composite."

pressed, the lower lungs are exhausted; the muscles passing from the lower end of the sternum to the pubis adjust themselves so that the lines of gravity of the body come upon the inner sides of the feet (reaction). The upper ribs being raised by their interchange of inner bearings *at their articulations* are filled (proper action). This for *secondary section*. For *primary section*, the axial line of filling for the upper lung lobes passes through the lower end of the sternum, at which, and following it, at the small of the back, the posterior cross-line ends adjust themselves on either side. Then the posterior points of application on each side at the anterior points of the pelvis.*

Bicomposite spine.—The hinder limbs interchange inner bearing (reaction). The shoulder-blades do the same (proper action) *secondary section*. The movement on the anterior ends of the head joint sockets followed by their movement backward on the outer edges of the head condyle, which changes the lines of general thrust and brings the bearing on the lower jaw, as representing the cross-line ends. Finally, the adjustment of the posterior points of application at the hip-joints

---

* At every stage more and more appui has been taken on the shoulder-blades.

in front, which are likewise for this spine, the continuation of the posterior cross-line ends, *primary section*.

On the *anterior lines*, (by induction)—*ophidian spine*. With movement of articulations of lower ribs on to their outer bearings, the lower lung lobes begin to fill for the *secondary* section. Then, with the movement of the anterior " points of application" at the lower part of the neck-rook joint, the adjustment of the anterior cross-line ends between the lower and upper lung-lobes for the *primary section*.

Composite spine.—With the continued movement of lower ribs on to their outer bearings, and continued filling of lower lung-lobes, comes the adjustment on outer bearing of the hip sockets for the primary section. Then the adjustment in continuation of the cross-line ends at the upper part of the sternum and at the small of the back* the anterior points of application rising in the neck above the neck-root joint for the primary section.

Bicomposite spine.—Adjustment of hinder-limbs on the outer bearing, the movement passing

---

* It will be noticed that the succession goes from sternum to small of back, with posterior lines leading, both for posterior and anterior lines—whereas it was from small of back to sternum, with anterior lines leading.

through hip joints for the *secondary* section. Then the adjustment of lower jaw articulations on outer bearing as representing the anterior cross-line ends, and movement of head condyles on their posterior ends, followed by movement of condyles across the joint and adjustment of shoulder-blades on the outer bearing for the final movement.

§ 183. We have rehearsed over again for double action equalization, the details which might have been collected from the movements on one set of lines. This has been done because the succession of double movements, with the posterior lines leading, appears to us to be that on which the *method* taught by Monsieur *Morquin*, before alluded to, is carried out. This method, which is excellent for a regular exercise,* we shall now discuss.

It will, we think, become evident from this exercise—

That although Mons. Morquin's method directed filling the lungs from the mouth, yet the normal filling of, is through the nostrils. That the upper lobes fill on the posterior lines, the lower lobes on the anterior lines. That the lower C C of the lung tips fill at the passage from

---

* That is for confirming the setting up by strengthening the deficient muscles. It does not, however, quite answer where a rapid "setting up" is called for.

posterior to anterior lines, when the "thorough action" of the head sockets passes the lines from the shoulder-blades to the lower jaw, and the upper C C of the lung tips on the passage from anterior to posterior lines, when the "thorough action" of the head condyles passes them from the lower jaw to the shoulder-blades. Also that each lung lobe, upper or lower, is exhausted and filled through its collateral nostril.

That when both upper and lower lobes are filled, the pressure from the lower lobes, if the filling of these through the nostrils be continued, may be made to drive out the air from the upper lung lobes, through the mouth, and that a long protracted, if not indefinitely continued, current of air may thus be kept up, on which the vocal chords may act, as the cords of any stringed instrument. Of the volume of sound created, and of the ease with which it may thus be produced, Mons. Morquin gave repeated proofs.

§ 184. Mons. Morquin, so far as we know, made no attempt at any anatomical explanation of the system, which, he said, had been taught him, as one of the soldiers of a battalion selected for gymnastic instruction, and which, at the time of the French Revolution of 1830 was stationed at Rheims.

Having been unable to find any reference to this method in the work of Col. Amoros, or to hear anything of it when in France, a number of years ago, we are inclined to think, that, as probably was the case with the motions of the manual of arms, the seat on horseback, &c., the method was derived from a description of the way in which some *perfectly formed* man *naturally* accomplished the action in filling his own chest and thus setting himself up. In this case, very probably, the instructor.

His directions may, we believe, although perhaps not in his own words, be correctly stated as follows :

(1.) " Place the feet parallel to each other, and together, throughout their length."

(2.) " Hold the head in its ordinary position, *and free*, but keep both it and the chest well forward, so as to bring the weight of the body on to the front part of the feet."

(3.) " Take a full breath through the mouth."

(4.) " Close the mouth, retaining the air thus taken in. Raise the chin only sufficiently to keep the balance of the body forward, without rising on the toes, and *push with the upper front part of the chest forward and upward*."

"*This action must, as the trial will show, cause the air to be expelled slowly through the nostrils.* This expulsion of the air is to be carried as far as possible."

(5.) "Retaining the position of all parts of the body, as thus attained, again fill the chest by the mouth, and again, by "pushing with the upper front part of the chest forward and upward," let air slowly pass out through the nostrils."

Remark.—"Let the inspiration through the mouth be sudden but full. Let the expiration through the nostrils be slow and regular."

"The repetition of these actions will at last bring the body into a position in which the chest will be tensely swollen with air, and every joint of the body feel free. Its attainment in the course of the movement will be marked by the feeling that the shoulder-blades are drawn firmly against the body, while at the same time the hands are turned, thumbs outward, elbows near the body, and the neck is perfectly free in front."

§ 185. *Explanation on the theories already advanced.* The whole difference between the above detailed method and No. 12—" Setting-up by double action, the posterior lines leading "—consists in the air for the upper lung lobes being drawn through the mouth in the former, which makes it possible *to begin* by filling them, and to follow this by exhausting the lower lobes by the "reaction" from the secondary section of the posterior lines, with the upper part of the chest as an intermediate starting point, instead of beginning at once with the shoulder-blades.

In No. 12 the *lower* lobes must be first exhausted through the nostrils, and then the upper lobes filled in the same way at each stage (excepting so far as the air already in the lungs may adjust itself). In Mons. Morquin's method the *upper* lobes are first filled by the mouth, and then the lower lobes exhausted by the nostrils. When the upper lobes are once filled, the lower lobes, in both methods, are filled through the nostrils, and the two methods coalesce. A continuance of filling the lower lobes beyond their capacity, will transfer the air to the upper lobes, whence it may be expired through the mouth, and form the voice.

Mons. Morquin's method has several advantages, especially that the filling of the upper lobes through the mouth, although not thorough, is rapid; and that, by a few mechanical directions, a recruit may be caused to go through a process which could hardly be explained to the mass of men.

In all methods commencing with the posterior lines, it is to be particularly remembered that no stiffening of the lines from the chest to the hip joints be allowed in the "reactionary" working of the lower ribs, because this traction passes backward only gradually, viz: first, by a change of gravity taking effect on the front part of the feet; second, by the adjustment in the muscles joining the lower end of the sternum and the front of the pelvis; third, by the drawing of the posterior points of application from the *front* of the hip joints (§ 162, sixth).

§ 186. By a movement analogous to that just given, but carried out on a single line, viz.: by filling the left upper lung lobe through the mouth, and emptying the left lower lung lobe through the left nostril, it will, we think, become apparent that Mons. Morquin's *method* is *a formation of the alternate lines*, and not a forcing of the old ones.

Let the left* upper lobe be filled from the mouth, the air being carried to its upper part, and then the left upper part of the chest being strained upward

---

* It need scarcely be repeated that all these exercises suppose the right-handed deformity. In the contrary case the parts mentioned would change.

and forward, let it be so carried around to the left and backward as to cause compression of the left lower lobe, such that the air passes out through the left nostril. After a certain amount of repetition of these actions the right lower lobe will begin to fill from the right nostril. After the right lower lobe has thus filled to a certain extent it will begin to compress the left lower lobe, and to form a co-working with it which may be likened to the working of the head condyles in their ball and socket connection. After this has been carried to a certain extent, a further enlargement of the left upper lobe will take place by transfer of air from the compressed left lower lobe. The left upper lobe will join in a sort of co-working with the right upper lobe, which, after compressing the latter and causing the transfer of air to the right lower lobe, will bring on a slight — finally adjusting — filling of the right upper, and conjoin the action of the two upper lobes, so that they may be likened to the *sockets* of the ball and socket of the head joint.

§ 187. It may be that the reason for the fact that few persons can take a so satisfactorily full inhalation through the nostrils as through the mouth consists in the necessity for a full movement of the lung tips, or, what comes to the same thing, of the "neck-root joint," in order to accomplish the first, whereas an inhalation through the mouth allows a greater filling of the upper lobes without moving the lung tips.

§ 188. A few directions given by Mons. Morquin for various exercises may be introduced here:

(A) Extend the arms and fingers to either side, the fingers being kept close together. Push with the ends of the fingers from the body, thus stretching the fingers to the utmost. Very soon one of the fingers will experience a tendency to close* and the others to follow it.

Keeping up the tension, let the fingers bend on their own joints (i. e., not at the knuckles) till the tips of the fingers rest on the inner faces of the knuckle joints.

Next close the knuckle-joints until the nails are supported against the ball of the thumb, and the heel of the palm. Thus the fingers are well supported. This exercise seems an excellent one for developing the strength of the *hand*.

Finally, bend the thumbs which, meanwhile have been kept tensely extended, so that their nail sec-

---

* We think the middle finger first. The stretching is the outer bearing carried to its extreme point—the bending the induced inner bearing.

tions rest against the central sections of the two first fingers.

(B) In *dropping from a height* the chest should be kept swollen with air, the upper part pushed forward and upward. The chin up, the feet close together, toes touching. The arms extended and stretched tensely upwards. The fingers together and stretched. Push upward with the finger ends.

On touching the ground, air should be allowed to pass out through the nostrils, and, if necessary, through the mouth with a shout. Possibly, also, a spring upward should take place. The effort to push up the chest and the ends of the fingers must be continued throughout.

(C) If, when dropping from a height, the elbows be bent and brought to the sides, and then be jerked backwards or the fore-arms thrust forward, a considerable *change as to the point of descent* may be effected.

(D) As a general rule in all gymnastic exercises, whether on the gymnastic bars or in the manual of the musket, &c., the fingers should be kept stretched when grasping, and *the "heel of the palm" should always strike first*, i. e., the stroke of the hand should, like that of the foot on the ground in progression, be made on the outer bearing.

In the second motion of "*charge bayonet*," if the left hand be held thus stretched out, the musket on striking the heel of the palm and falling in the direction of the knuckle-joints, will close the hand in spite of any effort to keep it extended.

§ 189. We may add:

(a) That the whole of the difficulty so generally experienced in "*support arms*" comes from the left upper lung-lobe not being properly filled with air, and from the consequent dropping forward on its inner bearing of the left shoulder-blade.

(b) The ball and socket action between the lung lobes being the pivot for all the movements of the body, the shoulder-blades, which, by their position, work directly over this pivot, should support each other directly, when the man is perfectly set-up; that is, the movement of an arm in fencing goes through its own shoulder-blade *directly* to the other shoulder-blade, as its appui, while the head in a plomb on its condyles regulates the *subordinate* ball and socket movements of the hips. The same would be the case in writing, &c. The shoulder-blades do not thus support each other where the (right) handed deformity is present, for there is then a *restrained* adjustment of the hips and head which interferes.

In walking, &c., the movements of the pelvis in its ball and socket action at the diaphragm is met by the action of the head, and then co-adjusts with the ball and socket action of the shoulder-blades; so that these last (with more or less movement of the arms) regulate *each* step. In the horse the suspension of the formation of the alternate anterior line in the bicomposite spine (A'^3) until after the grounding of the free fore-foot, in the trot, is necessary in order to maintain the collateral balance. The same thing occurs in the walk of a man, after (e. g.) the left foot has left the ground and the right foot come down, a moment's delay should take place before the old bicomposite gathering on the left shoulder is discharged, and it is the imperfect performance of this which causes the backward hitch of the right shoulder after the right foot has come down so generally seen.

(c) A well set-up man will experience no difficulty in carrying the musket in old way, i. e., balanced on the middle finger of the *left* hand, the stock supported in the hollow of the left shoulder.

Having thus gone into the details of the movements in the two identical actions of halting and setting-up, it may, we think, be added that they all follow on a continuous movement of holding up the head.* The attempt to do this will of itself indicate what line is to commence the winding line movements, and on this others will follow, if it be born in mind—

First.—That although in actual execution the formations of the ophidian, composite, and bicomposite spines must run into each other, yet there are three distinct stages, all of which give some movement at the head joint, viz.:

The ophidian spine, a general movement, as of a simple ball and socket.

The composite spine, a more definite movement of the division into two parts, accompanied with a full movement of the corresponding C of the neck S.

The bicomposite spine, a complete condyle and condyle-socket movement.

These movements must bring on the filling of the alternate lung lobes and exhaustion of the abnormally filled ones for each line, anterior and posterior; and, with the beginning of the composite spine movement, as finishing that of the composite spine, the action of the lung tips.

Second.—That the action of the posterior line

---

*If the movement is to begin with a posterior line, which virtually brings down the head in front, the front of the socket will rise as the head is raised.

b b' consists so much in the reflected secondary action, that the actual primary action occupies but little time, and also the direct secondary section is so retarded that the filling of the alternate upper lung lobe may be kept back until the formation of even the composite spine is well advanced.

The accommodating of the lower C convex of the neck to the reflected secondary action in that part, must be particularly remembered, and the inward turn of the filling *lower* C of the alternate lung tip, and outward turn of the exhausting *lower* C, the outward turn of the filling *upper* C and inward turn of the exhausting one.

Finally, in causing all the movements to succeed a continued development of the first, and, by induction of the succeeding lines in the head-joint, it must not be forgotten that *every thing goes forward toward the formation of the collateral tractions*, and that thus, whatever the movement, the head keeps steadily *up*.

To complete this subject, see particularly Appendix II, p. 290.

PART VI.

Riding.

§ 190. We have endeavored to show how seriously the inequality of action in symmetric parts, and the consequent, imperfect, and "shackly" movement of central points may interfere with personal locomotion.

Not less do these faults interfere with the seat on horseback, indeed the trouble here is still more serious, for the points of appui instead of being on an immoveable surface, which will await the adjustment of the body, as does the ground, have place on another body, which is continually in motion, and, if not met in time, are at once removed. Thus, if the seat bones of the rider do not move equally, one of them, generally the right, is left behind by the motion of the horse. Hence, that hanging back of the right shoulder so generally to be observed, even in passable horsemen. When this fault exists, it is only by making the knees, instead of the seat bones, the points of appui, that an action of the shoulder may make up for this loss of position, and it is in fact thus made up, with a sort of hop on the other seat-bone, for which the appui on the knee, or perhaps on the stirrup, gives the necessary freedom.

§ 191. The grand difference between riding and personal locomotion seems to us to exist in the fact that, although the rider sits over the lungs of the horse, the traction which he himself receives at the seat bones is not in the ophidian but in the composite spine. Consequently his first action in following is to combine the composite and ophidian spines by the action at the upper end of the sternum for progression, at the lower end for retrogression, or for a check in progression. This combination once made, the remainder of the movement is executed in the regular succession.

Inaugurating his own movement by a *leading* action in the ophidian S, as does a man on foot, will cause a discrepancy of motion between the rider and his horse sufficient to loosen the seat of the most perfectly formed man. A little practice, however, should soon overcome this difficulty, and, we think, it may be assumed not only that a thoroughly set up man may be at once taught to follow the motions of his horse, but also that he may at once be able to manage the animal, for he has only to inaugurate in his own body the movements he desires *in such a way that they may meet the horse's gathering at the proper time for forcing or for checking the latter,* and, if the horse be properly suppled, it will follow up the impressions received.

The *Setting-up on horseback* may, of course, be accomplished by any of the methods we have mentioned, but there are two among them which appear to be the most appropriate. Both begin in the composite spine; that on the "anterior winding line" would seem the best fitted to progression, that on the "posterior line" to retrogression, or to a check in progression. Both follow in the course of the "alternate line" methods Nos. 5 and 6, but begin at intermediate positions, and not at the "points of application"—that for the left anterior line commencing with the left shoulder-blade, that for the right posterior line with the right upper ribs.

§ 192. *For the anterior line.* While the chin is raised well up and carried forward the left shoulder blade is pressed forward and to the right, as it were, across and through the neck. The relative bearing of the shoulder-blade is that which it assumes in No. 5, at the conclusion of the movement of the alternate (right) condyle acrosss the head joint. In the present case it *first* causes the actions and equalization preceding this relative bearing to develope themselves—the right thigh rolls out—the right lower lung lobe fills—the right upper corner of the sternum reduces the left, the left sterno-mastoid turning outward in its lower portion—the right jaw articulation rises, as it were, over the left—the right head condyle turns on its posterior end—then moves across the joint to the front, and confirms the position of the left shoulder-blade.

Second.—The movement of the left shoulder-blade, " as it were, across the neck," is continued, and developes the equalizing of the posterior lines, bringing out the left posterior by induction. The left lower lung lobe is condensed as the "reactionary" movement from the secondary section equalizes the lower false ribs. The left upper lung lobe fills with the "proper" movement of this section—the left shoulder-blade is set in the left alternate posterior line—the right head socket presses upward against its condyle as the right upper lung lobe is reduced—passes forward across the head joint, sets the right jaw articulation as the lower C of the right lung tip condenses, and tightens the right sterno-mastoid by a drawing from below.

§ 193. It is to be remarked that in the progress of all the setting-up movements it may be necessary occasionally to stretch the central line of the body so that the advance made may distribute itself to the various parts and leave the leading point again in position to draw in its first connection. There is also a slight, final, "double action" movement necessary to complete the setting-up.

§ 194. For the *posterior line.* The relative bearing of the right upper ribs is that in which, after the passing of the right socket across the head joint (No. 6), the right articulation of the lower jaw and the drawing of the right sterno-mastoid would bring them. The actions and equalizations preceding this relative bearing are *first* developed—the left lower lung lobe is condensed, somewhat, as the "reactionary" movement from the secondary section of the forming left posterior (alternate) line equalizes the left false ribs—left upper lung lobe fills with the " proper " action—in the bicomposite spine the left shoulder-blade equalizes with the right—the right head joint socket presses up against its condyle, and the pressure crosses the joint to the front, following which the right lower jaw equalizes on its inner bearing, and the right sterno-mastoid draws.

Second.—The same carrying forward, and to the left, of the right upper part of the chest being continued, the upper lung lobes are fully equalized, then the lower ones; the right lower lung lobe filling, the right head condyle turns in connection with the right articulation of the lower jaw, and as it passes forward across, the joint draws on the left shoulder-blade. This finally sets by a drawing from below, and a slight movement of the posterior point of application in double action completes the setting-up.

§ 195. The last mentioned method § (194), as will be observed, carries out the rule to " bring the right shoulder forward." Both may be used for either progressive or retrogressive movement, because, as was said of all the methods, they directly, if not immediately, find that stage of the horse's actions which coincides with their requirements. But we think that, to recapitulate the general directions, to *raise and carry forward the chin, while the left shoulder-blade, on its outer bearing, is, as it were, carried to the front and right, across the neck, and the spine stretched at intervals, in following the movement, so as to bring its effects into the trunk, and allow the sides of the pelvis to equalize,* the whole ending by a spontaneous drawing from below on the right sterno-mastoid, and followed by a filling of both upper long lobes, will best suit for the forward movements of the horse.

Again, that the general directions to *carry the upper part of the right upper ribs forward and to the left, so as to diminish the protrusion of the left lower (false) ribs, by sinking the lower end of the sternum into their cavity, while at the same time an effort is made to* MOUNT AS IT WERE, *the body, by a backward movement of its upper part, over and upon the left shoulder-blade; the spine stretching, in following the movement, so as to bring its effects into the upper chest, above which, as the neck-root joint equalizes, the* HEAD JOINT IS LEFT FREE *to allow of and adjust itself to the movement*—the whole ending by a spontaneous drawing downward of the left shoulder-blade, and followed by a filling of both lower lung lobes will best suit the backward movements.

§ 196. Many books have been written on the seat on saddles, and on bits.*

As to the *seat*, it may vary somewhat with the "make" of the man, but well set-up men will have one *uniform* enough even for soldiers.

As to the *saddle*, no saddle can be contrived that will be a complete defence against an uneven seat, or careless packing and adjustment of the soldier's

---

* Major Dwyer's is one of the best and most interesting.

"effects." Against the latter a vigilant officer may provide; the former can be remedied only by a good setting-up, and the unsparing punishment of every trooper who does not maintain it.*

§ 197. As to *bits*, although a horse may be taught to check himself under a severe bit, as he would before a stone wall, its use can no more be called riding than stopping the animal in such a manner can be called halting him. A curb bit, with a high port, may, in most horses, force up the *upper* jaw, and thus prevent the head joint from closing as it must for the inner bearing and the spring forward.† It also, by the leverage of its branches, gives increased power to the usual way of opening the lower jaw, but it is deficient in *lateral* action, and to some extent, by making the lower jaw the *chief* "artificial ground" for motion, in place of the eyes, it is subject to the same objection which that favorite of the French army, the Duke of Orleans, made to the — for preliminary breaking, wonderful—sys-

---

* The relation of the knapsack to the foot soldier is the converse of that of the rider to the horse; if the man's shoulder-blades be *flat*, and his *step be even*, a *well* packed knapsack will hardly worry him.

† If in stopping a horse, a man on foot force the snaffle *upwards* into the mouth, so as to open it by the upper jaw, this action will, we think, be at once recognized; the direction to hold the hand high in "stand to horse" would seem to depend on the same principle.

tem of Baucher: "Je ne veux pas de système qui prend sur la vitesse des chevaux."

The double-jointed snaffle, conjoined with the other "aids," should, under a well set-up rider, control a horse reasonably well made, and which the man has ridden for a fortnight, under all circumstances; but then the other "aids" must often precede, and be only *met*, by that of the bit. In fact, the rider must imitate the motions in his own body, and his seat first communicate them to the horse.

§ 198. The "AIDS." Although a perfectly suppled horse will generally answer to the movements given by the body of his rider, yet if unsuppled, or fractious, certain forcing influences are required.

These are called "*aids*," and together with them we shall discuss some of the changes of gait and of action which they are calculated to produce, and which we have deferred from § 159.

§ 199. The aids are four in number, namely, the *Bit*, the *Spurs*, Pressure by the *Seat bones of the rider*, Pressure by the *Reins on the horse's neck*.

§ 200. The *Bit* has differing actions according to the changing relative bearings of the side of the lower jaw on which it acts. For example, if the left hind foot have just come to the ground, in the trot, the pressure of the bit on the left side of the mouth will increase the formation on the outer bearing, and if continued after the right posterior line has begun to develope, will more or less hinder the formation on the inner bearing which this requires. If applied exactly after the spring from the left hind-leg, it will hinder the completion of the alternate (right) anterior line in the bicomposite spine, § 111 (since it is the opposite side of the jaw which must then take an outer bearing), and so check the progression. Its effects, we suppose, may always be calculated for the "*working* side" of the jaw by its coincidence with, or opposition to the movements of the lower jaw, as these coincide with the winding line in progress of development; and, for the *unengaged* side, by the effect it thence produces in favoring or hindering the movements of the working side.

In double action, drawing on the bit favors all the anterior winding lines, and hinders all the posterior ones, excepting just at the interchange of condyles on the spring, when hindering the posterior winding lines checks the formation of the alternate anterior ones.*

---

* The movement of the jaw, it will be remembered, from its connection with the digastrics and their connection with the lungs, permeates the whole body.

§ 201. The *Spur*, and the pressure of the *Seat-bones* of the rider are so related that they must be discussed together. It is scarcely necessary to premise that we do not intend actual use of the spur when the pressure of the leg suffices. The spur, acting near the rear end of the sternum, brings on the action of the posterior winding line of its own side; thus the left spur will induce the developement of the *left* posterior winding line, first in the "reaction" from the secondary section, then, the "proper" secondary section, then in the primary section (the drawing of the cross line end). It thus reduces the working of the opposite posterior line.

If the left hind-foot have just grounded in the trot, the right posterior line begins to form and put the foot on its inner bearings; drawing the left rein will interfere with this, by checking the coinciding movement of the left lower jaw articulation on to its inner bearing, (temporal muscle setting); and the left spur will also check it by developing the opposite posterior winding-line; consequently the left hind leg, if (as in the right-handed man), it works too much on the inner bearing, will be "bent" or "suppled."

The Seat-bone pressure affects similar results for
12

the *anterior* winding line of its side, beginning, however, with the primary section. Thus, if the horse's right hind-foot be raised, the weight of the rider's body, thrown *perpendicularly* on the right seat-bone, will bring the right hind-foot to the ground on its outer bearing by the primary section of the right anterior line, without fully developing the secondary section. It also reduces the working of the secondary section of the right posterior line.

In ("bending") or "suppling" a hind leg, the right seat-bone pressure should work with the action of the left rein and left spur, so soon as the horse's left hind-foot has grounded. In this way the formation of the right posterior line is hindered by the left side of the bit, while the left posterior and right anterior lines, by their developement through the left spur and right seat-bone, subtract from the over-done "left-right counteractions" and tend to equalize the muscles.

The *spur in the flank* passes along the line on which it may be applied, from the posterior end of the sternum to the *back*, at which point it can be supposed to have a like influence to the seat-bone. Hence it may be that Abd el Kader described a perfect horseman as being able *croiser les eperons sur le dos de son cheval*, i. e., rowel him from the belly to the back at one sweep, which, no doubt, produces an effectual gathering.

§ 202. We have emphasized the word "*perpendicularly*" in speaking of throwing the rider's weight on to the right (or left) ribs of his horse, because we believe that few, even of pretty well made men, can do this. In general the weight is not *perpendicular*, even on the left side, and the attempt to pass it to the right side, being nothing more than a hanging over from the left, produces but litttle effect upon the horse. The required movement of the cross-line ends in the rider's body between the upper and lower lung lobes, is much greater when his seat bones are to interchange as appui, than what might serve tolerably well for the interchange of his feet; hence the difficulty. The value as an "aid" of this change of the rider's weight from one side to the other is little appreciated because few can use it.

§ 203. Pressure of the *rein* on the side of the neck. In our standard illustration for position — appuis of the horse on left hind and right forefeet — the left rein, if carried to the right, would press against the left convex of the lower C of the neck S, and would force it toward forming a convex to the right. Now, it has been said (§§ 60, 148) that in the regular change of curvatures the convex must pass *over* the concave; that is, it must, in the reduction, follow the direction of the line which formed it. This line was, in the present case, the secondary section of the left anterior line, whose convex would pass *over* and to the left. If, in so doing, it developed the corresponding section of the right anterior line, that convex passing to the right would oppose it. But it will, we think, be apparent that the pressure of the rein reduces only the lateral development of the convex, and, so far as the perpendicular development is concerned, has rather the contrary effect. Hence, *it will be the primary section of the alternate (right) posterior line which pressure of the left rein on the left convex of the lower C of the neck will develope.*

Again, the S being formed by the counteractions of two forces, compounded each of an element of pressure and an element of rotation, it may, we think, be assumed that the pressure elements give the longitudinal thrust, whereas the rotary element in each *secondary section gives an outward sideway movement to the ribs, legs,* and other parts dependent *on the convexity, whether this last be formed or only forming.*

Connected with a *primary section, the leg is car-*

ried inward across the body, for the movement there depends, not on the general course of the winding line, but on the direction in which the cross-line end is drawn by the point of application, and whether the leg, in connection with a convexity, is following the reduction movement of the old cross-line end, or, in connection with a concavity, the establishment of the new one, the direction is alike across the central line of the body.

The reactions will hold good for all the "spines" (§ 118), and thus the FORE-LEGS *may receive lateral* as well as other *motion from* two sources, *the body or the neck*, and be differently moved accordingly as they are in the trotting or the pacing connection.

§ 204. Since the spur (§ 199) developes the *primary* section of that posterior winding line, whose point of application lies on its own side — i. e., the left spur the primary section of the left posterior line, and so on — its application on a convexity would cause (§ 203) the corresponding hind-leg to move across the body with the reducing posterior cross-line end.

When the left rein pressed upon the left convex of the lower C of the neck, at the same time that the right bit, drawing the lower jaw of that side on to its outer bearing, checks the formation of the (alternate) left posterior line, the longitudinal thrust is suppressed, and, if the horse be kept steady, the lateral effect only has place. This would pass the left fore-foot in the neck connection across the body to the left, following the reduction of the old posterior cross-line end in the neck. Were the left foreleg, in its body connection — i. e., on the anterior concave — the primary section of the forming right anterior line would move it in the same direction.

§ 205. The *cavesson* having its action on the nasal bone, of course moves the *upper* jaw downward, and with a rein to each side from the projecting ring, the head joint may be influenced by closing it in front on either side, or using both reins on both sides at once. This closing of the joint in front is normally the result of the action of the posterior winding lines, and we should thus have an "aid" which would *directly* * act in favoring these lines as the bit does in favoring the anterior lines.

Possibly the rider, with a left cavesson and a left snaffle rein, and a right cavesson and right snaffle rein, crossed in either hand, might find the cavesson an additional aid in suppling his horse.

---

* The lower jaw movement for the posterior lines is one of closing — i. e., the inner bearing — for the anterior lines of opening — i. e., the outer bearing. Now, as the bit only opens the mouth, the posterior lines can only be favored one at a time with the bit, by the indirect action of causing one side of the jaw to close by opening the other.

§ 206. We subjoin two tables, the one giving the mode of action of the several aids, the other the lateral movements of the legs :

AIDS.

Direct.	Indirect.

Left BIT *favours* secondary section left anterior line,
    impedes primary section right posterior line.
CAVESSON. Right traction *favours* primary section right posterior,
    impedes secondary section left anterior line.
Right SPUR *favours* secondary section right posterior line,
    impedes primary section left anterior line.
Left SEAT-BONE *favours* primary section left anterior line,
    impedes secondary section left posterior line.
REIN. Pressure of right rein on lower C, left convex. *Brings on* left posterior line for neck S, and influences analogous convexes in the same way.
	Right BIT *favours* primary section of the right posterior line.

LATERAL MOVEMENTS OF LEGS.

Outward from central line.
Left fore-leg. Body movement. Secondary section of left posterior line.
    Neck movement. Secondary section of left anterior line.
Right fore-leg. Body movement. Secondary section of right posterior line.
    Neck movement. Secondary section of right anterior line.
Left hind-leg. Body movement. Secondary section left anterior line.
Right hind-leg. Body movement. Secondary section right anterior line.

Inward, across central line.
Left fore-leg. Body movement. Primary section of right anterior line.
    Neck movement.* Primary section of left posterior line.
Right fore-leg. Body movement. Primary section of left anterior line.
    Neck movement* Primary section of right posterior line.
Left hind-leg.* Body movement. Primary section left posterior line.
Right hind-leg.* Body movement. Primary section right posterior line.

---

* All these follow the direction of the old cross-line end, as it is reduced by the new one, e. g., the left hind-leg, that of the end belonging to the right posterior line, as the left spur developes the left posterior line. We have, for convenience, connected the motion with its primary rather than its proximate cause.

§ 207. No rider who cannot feel the manner in which his horse's feet are placed can accurately apply the aids. This is one of the decisive arguments for a close seat, without which such feeling is out of the question. A good seat once obtained, nothing is easier than to follow the advice of a German teacher, *Seeger*, and, knowing the sequence of the feet in the trot, to watch the fore-legs, and try to recognize by feeling what one knows to be the accompanying position of the hind-feet.

CHANGES OF ACTION.

§ 208. There are, of course, various ways in which the same changes of action may be accomplished. We shall endeavor to select for our explanations that one in which the horse would accomplish it under the influence of the rider, and, although we may occasionally differ from Von Oeynhausen, we must again repeat our acknowledgements to him for the "succession" of the legs in many, though not all, of the cases, without, however, at all charging him with our theories in regard to them.

Trot to walk. Supposing that in the trot the horse has just put down the diagonal right fore and left hind-feet. The left anterior line is not completed (§§ 130, 142, 143), that is, a^3 and a'^3, which turn the raised left fore-foot on its outer bearing and thrust it forward, are not yet carried out. With this completion, the left articulation of the lower jaw must come on its outer bearing. This the rider hinders by *drawing the right rein*. In the same moment, with the left spur, he temporarily and partially hinders the development of the right posterior line by commencing the formation of the left one. This hindrance to the completion of the left anterior line, and check in the formation of the right posterior, will induce the horse to change the working head condyle from left to right *directly*, i. e., without the intermediate movements, and in connection with the neck only.

The right head condyle will then raise the left fore-foot in its *neck connection* and put it down, whereupon the ophidian cycle, only suspended in its action, will resume the trotting movement for the hind-legs, the right hind-foot will be put down, and the walk inaugurated, § 145.

Gallop to Walk. In the preceding change of gait, "Trot to walk," the ophidian gathering for the trot step with the hind-leg was only restrained, but the fore-leg gathering was altered to the pacing action by changing the working head condyle. Supposing a horse in the "gallop to the right" to be halted for an instant as he lands from a spring. His feet are in position to step off with the left fore, followed by the right hind-foot, *if the working condyle be changed.* The horse, of course, could easily accomplish this, and possibly, the rider passing the bridle hand to the right, so as to develope the left posterior line in the neck by pressing out the convex, at the same time assisting this by a very slight action of the left spur, and immediately following the change of condyle by passing the weight to the right seat-bone, in order to put down the horse's right hind-foot, might teach him to do it.

Von Oeynhausen* remarks, "to change literally, at once, from the gallop to the walk, demands such precision on the part of the rider, in giving the aids, and such patient waiting for, and ready answering to them on the part of the horse, that it is hardly ever *really* done. In almost all cases, in common life, the horse takes a few short trotting steps, and then first begins actually to walk." In these steps the horse gradually eliminates a^2 a'^2 and b^2 b'^2 from the action (§ 145).

Gallop to Trot.—The horse being in gallop

---

* " *Gang des Pferdes und Sitz des Reiters* " plate 44 text.

to the right, it will be necessary to carry out the left right counteraction entirely through the neck-root joint, instead of allowing the right-left counteraction to join and form the double action. For this purpose, as the horse lands from a spring, the rider would weight strongly his left seat-bone in order to drive forward the left anterior line, use the right spur to strengthen the right posterior line, and, at the same time, give the horse his head sufficiently to allow him to respond by carrying through the trot, on the right fore and left hind-leg, landing on the left fore and right hind-feet.

Trot to Gallop.—For gallop to the right, the gathering for the left right counteraction having been made predominant, the right left counteraction must be introduced BEFORE *the completion* of one of the steps on the left hind and right fore-feet so as to bring on the double action (§ 156).

The rider shortens the right (inside) rein, carrying his hand to the left, which give a preponderance in working to the left head condyle; he also throws his weight on the left seat bone and uses the right spur; these aids develop preponderatingly the left right winding line. As the horse lands on the right fore and left hind-feet, he gives a sufficient amount of pressure to the right seat-bone, and sufficiently uses the left spur to introduce the right left winding line in subordinate connection with the left right, and thus forms the double action. Raising the bridle hand, he opens the mouth by the movement of the *upper* jaw, and by the necessarily following movement of the head condyles, which initiates the alternate anterior lines, and consequently the discharge of the spring.

§ 209. *Halting from the Gallop.*—Holding the reins steady as the horse lands, so as to check the motion, but not to change the head condyles by their movement, pressure with the right seat-bone brings out the right anterior line, and the left spur the left posterior. The development of those, the subordinate lines of counteraction restrains the left anterior and right posterior lines, and the head condyles not being allowed to change by their own movement, the four lines are equalized throughout the body, and the condyles conform to the new distribution of tractions.

§ 210. *Rearing* and *Kicking*—May be explained entirely by the "double trot actions."

Rearing.—In this the horse developes the anterior winding lines to an undue degree at the expense of the posterior lines. The hind-feet thus come extravagantly upon their outer bearings and the whole body is drawn back upon them as appuis.

Under the rider, a bit which prevents the horse from completing the primary section of the posterior lines when he is urged forward, may induce rearing. The lower jaw, checked in the attempt to come on its inner bearings, throws back the tractions to those which belong to the jaw on its outer bearings, i. e., the anterior lines in their secondary *sections*. If, when up, the horse thrust forward the front legs on the inner bearing by introducing the posterior line in the anterior C C, we have the full converse of kicking, which begins with the action of both lines in the posterior C C, and ends with the single action of the posterior lines in the anterior C C.

If now, the horse rear with a perfect equality of the sides, the constantly increasing action of the anterior lines will finally eventuate in "setting-up" by double action on these lines (§ 180), and he will come down perfectly gathered. But most horses, and—if they wish to resist their rider—all horses rear with a preponderating action of one leg, and if the rearing then be carried too far, they may fall over.

As a remedy for rearing, determined spurring,

by forcing the posterior winding lines to form, may bring the horse down from any position short of the loss of balance, but, as horses generally use a favorite hind-leg, the development of the counteracting lines for the other pair of diagonal legs will generally answer the purpose, if applied early enough in the movement. Thus, if a horse stiffen the left hind-leg, the rider should, by throwing his weight on the right seat-bone, bring the horse's appui more on the right hind-leg; at the same time (if necessary) lift the left hind-foot with the left spur.

§ 211. *Kicking.*—This is the converse of rearing; the horse developes the *posterior* winding lines to an undue degree at the expense of the anterior lines; the *hind-feet* rise on their inner bearings with a forward movement, and are next thrust out to the rear by a backing movement, as the *fore-feet* come on to their outer bearings, by reason of the *unmixed* action of the secondary section of the posterior lines in the anterior C C.

In kicking, as in rearing, most horses have a favorite leg; supposing this to be the left hind-leg, appuied on the right fore, then the left spur, followed by the weight on the right seat-bone, which would introduce the *alternate* right-left counteraction, should equalize the lines and reduce the unmanageableness of the kick.

§ 212. *Bucking.*—This might, we think, be explained as a jump *upward* in the double pace movement, while refusing the cross-line action of the spine. The thrust in the spine for an upward jump is both ways from the cross-lines as a centre.

§ 213. *Turning.* The common turns in the trot we should describe as being brought about in the following manner: The drawing of the rein on the side toward which the turn is to be made can be done under two conditions, which give origin to two very different steps.

First.—When, for example, the right fore and left hind-feet have just landed. The neck portion (a^3 a'^3) of the left anterior line—completing the bicomposite spine — is about to form, followed instantly (or possibly somewhat preceded) by the right posterior line. Now, since the formation of the left anterior, followed by that of the right posterior line, will bring the *left* articulation of the lower jaw on its inner bearing, (i. e., the left temporal muscle drawing) the pressure from the *right* side of the bit, by drawing the RIGHT rein favors this; but, at the same time, it hinders the thrusting element of the two lines, and, so far as this goes, the primary section of the right posterior line crosses the right (free) hind-leg to the left. The right hind-foot being put down, the alternate right-left line forms and the primary section of the right anterior line passes the left fore-foot across to the right.*

The whole movement may be strengthened by the right spur, increasing the working of the right posterior line.

If the horse be in progressive motion, the right hind-leg will, we think, be found the first to reach the ground; if he be stationary, he will *back somewhat* on the right hind-foot, thus holding that part of the left-right counteracting lines undischarged, while the alternate right anterior moves the forefoot, and, as it developes, discharges the old line (§ 171), which last crosses the right hind-leg.

Second.—When the right fore and left hind-feet having just landed, the LEFT rein is drawn. As in the previous paragraph, a^s and a'^s are about forming in the bicomposite spine, but pressure of the bit on the *left* side of the jaw will hinder the

---

* These crossings are thus both *body* movements, the left fore-leg being forced back under the influence of the front C of the ribs. The distinction between this, which is a trotting movement, and the crossing of the fore-leg in " passage" (§214), which is a pacing movement, will be noticed.

formation of the right posterior line, and the drawing on the head suppress a^8 a'^3 by causing the horse to change the working condyle;* consequently, the thrusting element being suppressed, the left fore-leg will be darted to the left by the rotary element of the secondary section of the right anterior line *in the neck*, and the right hind-leg will follow with an outward step caused by the rotary element of the secondary section of the right anterior line *in the body*.

These two stages will exhibit the working of the bit for two or more steps in the same change of direction; for, when turning to the right, the *second* action of the bit occurs for the left fore and right hind-legs as appuis, the *first* for the right fore and left hind. Thus, at one step, the horse crosses the free legs, at the next, he throws them outward from his body.

There is still another way of changing direction, the discussion of which must be reserved for "circling on the haunches" (§ 217), of which the movement is simply modified by progression being more or less continued as it proceeds.

---

* It should be remembered, that in the horse the head follows the lower jaw only when the latter is closed; when open the lateral movement disengages it.

§ 214. *Passage.*—So-called in the United States and in the English Cavalry Tactics—(French, *Appui*—German, *Schliessen, Half* and *Full Travers*).*

Taking from the tables §§ 206, 207, the rules that the rotary elements of the *anterior lines* in each of the three "spines" act by their secondary sections in connection with the lower jaw on its outer bearings, to throw the limbs outward from the central line of the body; while the *posterior lines* act with the lower jaw on its inner bearing by their primary sections, to throw them inward across the central line, we should explain the "Passage" in the following way:

The horse is placed as if for progressive movement on a pair of diagonal appuis—say on the right fore and left hind-legs, the head is then confined by drawing the right rein, so that the (working) left head condyle cannot actually discharge. The left spur, in the next place, bringing into action the left posterior line, raises the left hind-foot and forces down its right fellow. The formation of this line would be a part of movement forward on the right hind-foot, which its complement the right anterior line not being formed, the

---

* The "Passage" proper is not the same, but a sort of "Mark-time" in the trot.

horse might shirk by backing on this foot—but the rider by bringing down his right seat-bone and partly inducing the right anterior line prevents it. Now, the action of the left spur forming the left posterior line up to its primary section, should, in connection with the pressure of the right side of the bit which checks its thrust, carry the left hind-foot to the right, across the central line of the body. The pressure of the left rein carried against the neck convex, while tending to produce the alternate curve in its left posterior line component, should, by the primary section of this line, carry the left fore-foot, related to the neck as the left hind-foot is to the body, in the same direction, viz., to the right—and this pressure on the neck affecting the whole length of the spine, forces the horse, if he have resisted, to yield to the foregoing action of the left spur.

Under the actions of the left spur, left rein and right seat-bone, the alternate winding lines have nearly suppressed the left-right counteraction, but the head condyles have not been allowed to change. This counteraction is now restored, and being still held in check, as to its forward thrust, by the *left* side of the bit, the rotary elements of the secondary section of the anterior line, on the lightening of the right seat-bone pressure, carry the right fore-leg with the neck restoration, and the right hind-leg with that of the body sideways to the right, by reaction from what would have been the movement of the two left feet had they been free.

The neck action in this movement connects it with the pace on both sides of the body.*

The horse resists the "passage" from a halt by backing, and, when in progression, by striking a pace with the (right) "inside" feet.† The former is checked by the pressure of the (right) "inner" seat-bone of the rider, and for the latter, the inside rein must be drawn sufficiently to prevent the change of condyle.

The "passage" is one of the best exercises for suppling, particularly when the horse has a "favorite" side of the mouth for resisting the bit, and it is also the best remedy for shying. For the latter, the horse should be made to passage *toward* the object which he avoids. In resisting this, he will very possibly strike a pace which, we think, cor-

---

* The *diagonal* legs in the turn (§213 second) were thrown outward by similar but not the same movements, for there they occurred on two different lines of counteraction, these on one and the same, as in the pace.

† Would be such if on the circle and passaging toward the centre.

roborates the view we have taken of the nature of the action.

§ 215. *Circling on the Fore-hand,* and *Circling on the Haunches.**—These movements, including, of course, the pirouette renversée, and the pirouette, seem to be both contained in the actions of the "passage." Circling on the fore-hand being the haunch movement, with the neck movement reduced to a minimum, and circling on the haunches the neck movement, with the haunch movement reduced to a minimum.

Circling on the Fore-hand, with the head turned inwards. The horse is put in position with the appui, say, on the left hind and right fore-feet. The rider lifts the left hind-foot with the left spur, presses down the right hind-foot with his right seat-bone, and continues the action of the left spur until the formation of the primary section of the left posterior line; the thrusting element held in check, carries the left hind-leg across the body to the right. *No pressure* being made with the left rein, as is done in the "passage," the weight is thrown upon the left fore-foot without moving it,

---

* These movements are well shown in the plates accompanying the late General Kenner Garrard's Annotations on Nolan, Baucher and Rarey.

and when, the pressure of the right seat-bone being lightened, the restoration of the left anterior line occurs, this, while in the *body* connection it passes the right hind-foot well to the right, in the neck connection only moves the right fore-foot sufficiently around its left fellow to readjust the position.

In the *pirouette* renversée, we should suppose that the addition made to the above movement was, that the horse somewhat increases the neck gathering, and retains it until the arc is completed. He raises and passes across the left hind-foot, and *springs from the right hind*-foot by the right anterior, left posterior lines maintaining, however, the left-right reaction by keeping the left head condyle in place as the working one. This last condition enables him with the right fore-foot to bear off the weight on to the left fore-foot. He finally descends on the left hind-foot, and then plants the right hind and right fore.

§ 216. *Circling on the Haunches.*—The horse is put in position, say with appui on the right fore and left hind-leg. The right rein is well drawn, so as to fully develop the lower C of the neck S, convex to the left. Then, with the rider's right seat-bone developing somewhat the right anterior winding line, the left rein is pressed against the neck by carrying the bridle hand to the right, and the left fore-leg forced across to the right. The left spur is used just sufficiently to start the movement to keep the right hind-foot a little on its inner bearing, and to insure the small required movement of the left hind-foot as it moves around its right fellow for a pivot, and then sustains the extended adjusting movement of the right fore-leg. The sideway movement of the right hind-leg to the right is represented only by its adjustment.

For the *pirouette* the right spur, resisted by the left side of the bit, develops the secondary section of the left anterior line, and the horse rises on the left hind-leg (§ 210). Then the pressure of the rider's right seat-bone, the pressure of the left rein, and, if required, the left spur, cause the horse, in the effort to carry the left fore-leg across, to face about on the right hind-foot as a pivot. The right spur keeps the horse from discharging the left head condyle, and, with the left bit, keeps him up. The rein pressure forces him around, and the left spur brings him sufficiently on the *inner* bearing of the right hind-foot. The seat-bone pressure must be delicately adjusted, as after raising him with that of the left, the right gives proper outer bearing to the right hind-foot, which may be said to be continually *corrected* by the inner bearing.

§ 217. *Changing direction on the inner hind-foot by pressing the outer rein* we should consider as circling on the haunches combined with progression.

§ 218. We have only to add, in conclusion to this part, that as the snake's motion has been taken as a clue in tracing *up* the mechanism of locomotion in the higher animals, so the horseman cannot, we think, do better than to reverse the process, and take the working of analogical parts of his own body as a clue for guiding his observations and conclusions as to the actions which take place in the body of his horse, and as to the best means of controlling them.

APPENDIX I.

LOCOMOTION OF BIRDS AND OF FISHES.

We have little to say concerning the locomotion of these classes of animals. It would seem to us that, although the fins of a *fish* are added, apparently somewhat in the manner of limbs, yet the fundamental locomotive action of a true fish comes from the tail, following the ophidian motions of the back bone, and acting on the water as the blade of a single oar does when worked at the stern of a small boat in the motion called "sculling."

Birds, we should say, fly by alternately raising and pressing down the front edges of their wings. The first motion presents the wing as a plane inclined upward to the air in the front-rear direction. This is the *outer bearing*, and on this plane they rise after the second motion, which is a downward stroke of the anterior edge, answering to the *inner bearing*. If the action of flying come fully under our theory of locomotion, the second motion should collect the feathers in a spring before its discharge.

Birds, as is commonly known, have no effective movement in the vertebræ of the body whilst the numerous vertebræ of the neck are very moveable. We should explain the S S actions of the neck in the following manner : Supposing the body vertebræ of a man to be thus solidified, there would be one of the motive connections of the arms, viz., that with the body, unprovided with a diagonal counteracting basis. This basis, as it exists, brings the legs and arms into connection, and both are then brought into a central line by the neck S. Possibly the extra S in the neck of birds (one or more above the number in quadrupeds and man) (§ 85) supplies this loss, and there is still the same double action on a bird's wing as on the arms of a man.

A bird cannot, we think, keep its head steady when walking on the ground, without stretching the neck. May it be that, the body, being then confined to one plane, and the lower S of the neck having no means of adjusting the excentric movements of the neck-root, these movements must be communicated to the head. If the neck be stretched, the action of this S is reduced as much as possible.

The spring collected in the wing and the double

action above alluded to, being absent in artificial wings, may have something to do with the poor success attending all attempts to adapt them to the human frame.

Possibly the reversed positions of the head and sockets of the rib articulations in the snake to those in the higher animals, may be accounted for by the discharge in the S S, being successive, for the former, but combined for the latter.

APPENDIX II.

We will attempt a concise general outline of Setting-up for the right hand deformity, bringing in a portion of the movement which has not been made sufficiently prominent in the previous descriptions.

First.—The (alternate) left posterior line leading. The head being continuously raised forward, the right upper ribs at the neck-root joint are pressed to the left and somewhat forward. This latter action, which begins a reduction of the right upper lung lobe in the ophidian S, on the course of the alternate posterior line leading, extends to the neck and to the left anterior part of the head-joint, when there the socket begins to draw away from the condyle.

Presently the left upper lung lobe begins to fill and to cross the secondary section of the (alternate) left posterior line with the corresponding portion of the (old) right posterior line. This crossing passes from the ophidian to the neck S, and causes the anterior part of the right head joint socket to rise against the corresponding part of the right head condyle.

A repetition of the foregoing movements soon causes the right lower lung lobe to commence filling, not on the secondary sections of the right anterior line, but—and this is the point which we wish to make prominent—on the *reflected action* for the (alternate) left posterior line.

This is continued until the left lower lung lobe is equalized by exhausting it on the reflected action of the (old) right posterior line, carrying the movement through the left hip joint. Next the resumption of the *direct action* on the (alternate) left posterior line completely fills the left upper lung lobe and fully reduces the right lobe, ending by the proper adjustment of the left anterior part of the head joint socket and the drawing on the right shoulder-blade from above.

A slight continuation of the movement will next bring on, by induction, the full filling of the right lower lung lobe on the (alternate) right anterior line, the consequent reduction of the left lower lung lobe, and, finally, the additional reduction of the right upper lung lobe and filling of the left upper, which all depend on the equalization of the two anterior lines. The whole ends with a drawing on the left shoulder-blade from below.

Referring the two fillings of the lower lobe, i. e., one from the reflected action of the secondary section of the posterior line, and the other from the secondary section of the anterior line, to double setting-up, it may be seen that the former causes no tightening of the muscles from the lower end of the sternum to the pubis bones, while the latter does. Hence, whenever beginning on one or both posterior lines, and, of course, in Mons. Morquin's method, this drawing is to be avoided until the ending of the movement.

It may be easily seen, we think, in what manner the formations on the anterior lines joining with those of the posterior complete the double twist of the ribs, etc., and thus join the filling of the lung lobes into one.

Second.—The (alternate) right anterior line leading. Here the head is carried up and forward, but, not drawing on the left sterno-mastoid muscle, passes with the neck somewhat to the right. It may need a slight pressure of the left shoulder blade forward and to the right in order to initiate the movement on the (alternate) right anterior line by which the right lower lung lobe commences to fill. This filling begun, the movement extends through neck and to the right head condyle, which begins to press on its socket. The reduction of the left lower lung lobe next begins on the (old) left anterior line, and the action extends to the neck, causing a marked movement in reducing the left convex of its lower C, and then in reduction of the left head condyle; both equalizing the (old) left anterior line with the (alternate) right anterior, crossing them at their cutting points in the three spines, and, from the left shoulder-blade throwing the tractions forward on to the whole right side line of the sternum and on to the right articulation of the lower jaw, and at the same time giving such equalization to the upper lung lobes as belongs to the anterior lines.

Finally, thrown back from the left articulation of the lower jaw, begins the filling of the left upper lung lobe in the direct, and of the right lower in the reflected course of the (alternate) left posterior line and the corresponding reduction of the left lower and right upper lobes, ending with a drawing along the right side of the sternum concentrated at its right upper corner.

It will be noticed that in beginning with the posterior line the final movement was the straightening of the spine to the right; as in the ophidian S, the drawing on the posterior cross-line end of the (alternate) *left* posterior line was established; while in beginning with the anterior line the final movement was the straightening the tractions of the breast bone, also to the right, as the drawing, likewise in the ophidian spine, on the anterior cross-line end of the (alternate) right anterior line was confirmed.

It was said that a theory would be proposed as to the course of the optic nerves; but in copying the last manuscript from the older one this was omitted.

It amounted simply to this:

The optic nerves, after leaving the back of either eye, run together at about a right angle. Some of the fibres cross each other, and some of them, it is supposed, continue on their own side. May it not be that the crossing ones go with the anterior winding-line of their respective nerves, and those which do not cross with its counteracting posterior line? Thus the crossing fibres of the right optic nerve would go with the anterior left winding-line, and those which keep their side with the posterior right winding-line, in all the spines.

INDEX.

	PAGE.
aa' In snake's movement	46-55
aa¹ Resume'	61
Abd el Kader—His definition of a horseman	262
Action, changes of in horse	270
Aids—Table of their action	268
" The four	259
Air—In forming voice	239
" Passes collaterally	117
" In chest assists gathering and discharge	59
Alternate lines	186
Appui, how secured at one end instead of centre	33
" additional in front	35
Artificial ground	126-128
" "	188
bb¹ In snake	47-55
bb' Resume of action	63
Ball and Socket Action at four points	120
" " " where	127
Bevels on ground ends of snake's ribs	50-51
Bearers—A division of the ribs	58
Bearings—When given off point recovers in opposite direction	191
Bicomposite Spine—Its bracing collateral	95
Bicomposite Spine	84
Birds, Locomotion of	287
Bit—Its action	259
Bits	258
Body—Only attachment to head in front of head joint is by the digastrics	126
Bucking	277
C Half a torsion curve	23
C Rear the first discharged by alternate anterior line	48
C C Posterior part the longest	56
Canter	172
" Its analogies with the walk	174
" How it becomes a run	177
Cavesson	260

	PAGE.
Centre of force between lung-lobes	136-188
Chest, Varying elasticity of different parts of	89
Chin must not be allowed to interfere by dropping	
Changes of action in horse	270
Circling—On the fore hand	283
" On the haunches	284
Cord A. May be twisted so as to resemble locomotive lines of the snake	22
Convex in discharge passes over concave	57
Convexes are the normal sides of appui	51
Collateral tractions. Become such when	71-120
Collar bone in man	99
Convexities fused into two when spines are combined (note)	169
Composite spine consists of	79
" "	83
" " Its bracing diagonal	95
Condyles, Head	88
Concaves—Their extension into and reversal of one another	190
Cross-lines	26-27
" " All plane sections between changes of curvature may be considered as such	39
Cross-line end, Lower Displacement of	34
Cross-line—Directions in which its ends are moved	31
Cross-lines—Their seats for the different spines	188
Curvature, Gradations of	30
Deformed movement, Cause of	78
Deformity—The right hand gives a sort of canter to step	181
Displacement of lower cross-line end	34
Discharge of spring caused by cross cutting of winding lines	53
" for posterior CC ascends, and for anterior CC descends	55
" of spring	153-154
Digits—How numbered	101
Dishing of fore-feet in horse	106
Diaphragm	108
Direct action of secondary section in posterior lines	117
Disunited gallop	178
Digastrics	112
" Their analogy with the diaphragm, etc.	121
" Their attachments	124
" Their working	126
" Correspondence with ball and socket action	138
Diagonal action	127
Double action	167

INDEX.

	PAGE.
Double pace (full run).............................	171
" trot...	177
Dropping from a height..............................	246
Eight, Figure of, Shape	56
Elements of torsion, Two............................	24
" of motion projected on base of skull	78
Epiglottis...	116
" Retains compressed air in lungs..............	117
Equalizations may begin at any point	185
" See setting-up...............................	
Eyes lead locomotion.................................	73
" How steadied.................................	73
" Pulley muscle of.........................74-112-128	
" Respectively pivots of diagonal rear appui.....	77
" The foot ends of all locomotion................	142
" In the trot...................................	142
Eye-muscles—How affected by anterior lines...........	183
Exercises given by Mons. Morquin.....................	245
French—A French system of setting-up.......... 239-238-240	
Feeling the movements of horse.......................	270
Filling of lung lobes and lung tips................	131-132
Final action in equalizing, the same as the commencing one	184
Fish, Locomotion of..................................	287
Fingers Represent ribs...............................	100
Flying, Reasons against artificial	289
Forcing the winding lines without discharging them......	68
Fore-foot of horse—Why it remains presenting for inner bearing after the lift.............................	135
Fore-leg—When substituted for digastric in connection with diagonal hind leg..........................	139
Focus of force the centre between lung lobes..............	188
Foot of horse—A theory of its analogies.................	105
Foot—How the whole foot is kept on ground while CC succeed each other in action..	103
" How constituted in man, horse, dog, etc.............	104
" Raised hind—When only it can be fairly put down....	137-138
Gaits of horse, Difference between	143
Gallop Change to walk.................................	271
" " " trot...................................	272
Giraffe—Movement of lower jaw........................	156
Ground, Artificial or real, What for each spine	188
Hand—Heel of palm should always strike first............	246
Halt, the (see Equalization Setting-up).....	165
Halting, three ways of................................	166
" Final action same as commencing one	184

	PAGE.
Halting, From the gallop	274
" How differing from locomotion	72
Head joint in snake—Its action	37
" " Relations of its different parts to the winding lines	91
" " How altered in the higher animals	87
Head—To be kept always raised up	250
Higher animals	76
Horse—Muscles answering to sterno-mastoids and a clue to action of the latter, 85 note	96
Illustrations for diagonal movements; always suppose left anterior and right posterior winding lines to commence the action	25
Joint Virtual, at roof of neck	89
Kicking	276
Lateral pressures at head joint	92-93
Larynx	116
Landing from spring—Order in which snake's ribs should come to ground	58
Legs Free—How moved in discharge	121
" Diagonal—Their movement more synchronous in retrogression	165
" Table of their lateral movements	269
Limbs, Fore and hind, Difference between	97
" Fore—How guided by trunk, and how by neck	133
Lines, Winding	28
" Alternate may lead in equalizing	186
Line of general pressure, Where oblique requires rounding of condyles	91
Lobes of lungs	112
Locomotion of man and of the horse	130
Loops of windpipe, When they allow lower jaw to close	125
Lower jaw—Its articulations	125-129
" " Artificial ground for bicomposite spine	127
" " Yields in retrogression	165
" " Goes especially with posterior CC	212
" " Analogues to its motion	97
Lungs in the higher animals	107
" Cells and tubes of	116
" Exercises founded on their action	182
" In the snake	74
" Their centre the focus of force and centre of setting-up	136-188
Lung lobes	112
" " Their peculiar shape	114

	PAGE.

Lung lobes, Manner of filling 114-131-132
    "   "   How their filling and exhausting accompanies
            the different lines of torsion 117
    "   "   Filling of each brings its C forward............ 190
    "   "   Filled through collateral nostrils............... 238
    "   "   Ball and socket action between them the pivot
            of all movement........................ 247
Lung tips 112
    "   "   Manner of filling........................ 115-131-132
    "   "   Analogous action in lower lobes.............. 115
    "   "   Their action................................. 238
Morquin, Mons—His account of himself 239
    "   His method of setting-up 238-240
Neck—Snake must use two or three vertebræ as such...... 60
    "   Its connection with the fore limbs................. 98
Neck-root—Its virtual ball and socket joint............... 89-90-128
Neck-root joint—On it the fore limbs change from their body
          to their neck connection........................ 98
Nostrils—Normal filling of lungs has place through them.. 238
    "   Why sometimes difficult to breathe entirely
          through them 244
Oeynhausen, von Colonel............................ 161-172-179-272
Ophidian S. 83
Orleans, Duke of—His judgment on Baucher's system..... 258
Passage.. 280
Pace... 154-155-156-157
    "   Details of................................... 158-159-160
Pelvis.. 78
    "   Muscular connection with breast bone................ 87
    "   Acts with both CC...................................... 142
Perpendicular plane, Movement in....................... 58
Pillars of diaphragm.................................... 110
Pirouette and pirouette renversée...................... 284
Points of application.................................... 25
Posterior winding line—Its working not completed until
          at partial discharge of spring.................... 137
    "   point of application—Its direct drawing almost at
          last moment in equalization..................... 192
Posterior winding lines—Reflex and direct action of their
          secondary sections............................. 117
Primary sections of winding lines....................... 29
Pressure of socket against condyle on one side depends on
          its separation on the other...................... 39
    "   Lateral at head joint........................... 92-93
Propellers—A division of the ribs....................... 58

	PAGE.
Progression—From the position of ready to discharge	65
Progressive locomotion, how secured	184
Psoæ muscles	111
Pubis bones—A continuation of the breast bone	79
Raabe, Capt.—His theory of the walk in man (note)	103
Reflected action of secondary section of posterior lines	117
Retrogression	45-65-163-164
Resume' of snake's movement	61
" of chief points in locomotion and halting	182
Results which may ensue on position of readiness to discharge	64
Rein, Pressure of on side of neck	263-286
Rearing	274
Rib collects spring	44
Ribs of snake	42
" " How double twisted	43
Ribs Slip at articulations	45
" How correspond with movemements of head joint	46
" How affected as to their bearings by the winding lines	51-52-53
" Respective rôles of those on the anterior and posterior part of a convex	56
" On concaves do not change facings until the spinal curves change	37
" Proportion of in higher animals	76
" True and false	80
" Proportion of false to true in several animals	81
" How assigned to the CC	82
Riding—How different from personal locomotion	252
" Why rider leaves his horse	251
"*Right shoulder forward*," in riding agrees with	256
S Least number of vertebrae which can constitute	84
Definition of	23
SSS Nomenclature	83
" Why three are required in higher animals	82
Sacrum	79
Saddles	257
Scutæ of snake—Their action	60
Sections of winding lines, primary and secondary	29
Seeger, Herr	270
Seat of rider over lungs of horse	252
Seat bone pressure—Its action	261
Setting-up	182
" " Tables of on the ophidian action	194
" " Remarks	206

INDEX.

	PAGE.
Setting up Pacing movement	211
" " On horse-back	253
" " Following a continued raising of the head	248-290
" " Appendix II	290
" " On the alternate lines, ophidian movement	214-217
" " On the alternate lines, pacing movement	220-221
" " Practical Ophidian movement	222-223-224
" " " Pacing movement	225-226
" " " Alternate lines	227-228-229
" " By double action Anterior lines	231
" " " " " Posterior lines	235
Shoulder blades, Act with	142
" " Their connection with the neck	98
" " Straightening of the figure concentrated between them	188
" " Go in connection with the trunk with anterior CC	212
" " Their effect on the winding lines when they are pressed directly on the base of the neck	230
" " Should work directly on each other as appuis	247
Snake—Why possibly its manner of rib articulation is reversed in the higher animals	289
" Anatomy of	36
" Its movement	40-61
" Probably incapable of locomotion by "double action"	49
Spine—Its mode of action	22
" Of snake, its anatomy	41
" Composite consists of	79
" Bicomposite	84
Spines—How the three combine	189
Spur—Its action	261-262
Sterno-mastoid muscles	85
Sternum—Intermediate appui when action of the neck leads	99
" Its movements	140
Stopping a horse by forcing bit upward (note)	258
Superimposition of twists, Definition of	67
Symbols used in explanations	144-168
Tables for equalization of the four tractions	194
Table for action of the aids	268
" for lateral movements of the legs	269
Temporal muscles	125
Torsions, Counter, reverse each other or themselves	31

	PAGE.
Torsion may be resolved into two elements	24
Toes Represent ribs	100
Toe, Great, Should spread inward	102
Transformation of action of posterior lines	39
Trot—General description	130
" Details of A	145-148
" " of B	148-153
" Change to walk	270
" " to gallop	273
Turning	277
Twists—Discussion of theory	25
" Superimposition of Definition	67
Vocal chords	239
Vertebrae in the higher animals	76
Walk in the horse	161-162-163
Winding lines	28
" " Anterior and posterior traced for composite spine	161
" " How posterior becomes alternate anterior	179-180
" " How they affect the rib bearings	183
" " Their general course in the bicomposite spine	186
" " Their sections	92
" " Nomenclature	30
Windpipe—Artificial ground for neck action	126
" How suspended	121
" Torsion and counter torsion in	123

www.ingramcontent.com/pod-product-compliance
Lightning Source LLC
Chambersburg PA
CBHW031906220426
43663CB00006B/795